200% of Nothing

An Eye-Opening Tour through the Twists and Turns of Math Abuse and Innumeracy

A. K. DEWDNEY

John Wiley & Sons, Inc.

New York • Chichester • Brisbane • Toronto • Singapore

Library of Congress Cataloging-in-Publication Data

Dewdney, A. K.
200% of nothing : an eye-opening tour through the twists and turns of math abuse and innumeracy
A. K. Dewdney
p. cm.
ISBN 0-471-57776-6
1. Mathematics—Popular works. I. Title II. Title : 200% of nothing. III. Title: 200% of nothing.
QA93.D49 1993
510—dc20 92-42173

Printed in the United States of America
10 9 8 7 6 5 4

Dedicated with affection & respect
to
William H. (Pop) Adamson

The urgency of X
outweighed 33 teenagers.

Preface

What if you are already a mathematician and don't know it? Near the end of this book you will learn that we all use logic (of a largely unconscious kind) to take us through everything from business meetings to family dinners. If only we could be good *conscious* mathematicians!

At the conscious level, too many of us are still innumerate. We cannot deal with fractions, large numbers, percentages, and other relatively simple mathematical concepts. Meanwhile, we are beset more than ever by the abuses that stem from this innumeracy. And those who abuse mathematics also abuse us. We become prey to commercial chicanery, financial foolery, medical quackery, and numerical terrorism from pressure groups, all because we are unable (or unwilling) to think clearly for a few moments. In almost every case, the mathematics involved is something we learned, or should have learned, in elementary school or high school.

The first part of this book illustrates the most typical abuses of mathematics with examples drawn from the real world. Little gems of misunderstanding—stark, blatant, and grimly enjoyable—they amount to a primitive typology of mathematical crime. Each one illustrates how a particular branch of mathematics is warped by personal beliefs, public agendas, or corporate goals.

The second part of the book looks through the other end of the telescope, so to speak. It examines the areas of everyday life where abuse is especially rampant: the media, gambling, politics, finance, commerce, and, of course, advertising. Each area suffers its own pattern of abuse: horror stories of mangled percentages, tales from the crypt of compound growth, and so on. Readers who become alert to the underlying patterns will be well on their way to functional numeracy. Some, I hope, will go even further. There are unexpected treasures in the nearest university, community college, or night school.

Almost all the examples that appear in this book were sent in by readers of my columns in *Scientific American* and other publications. Mostly recent, they amount to a cross-section of what is actually going on out there. In a world of declining educational standards, oportunities for innumeracy and abuse proliferate.

No single book can stem the tide of abuse. But, in the third and final section, this one examines both the broad, long-term issues and the ones immediately at hand. Our educational system has entered a profound crisis. A faltering education system is colliding head-on with the worldwide emergence of a competitive technological economy in which mathematics, more than any single science, will be the main determinant of excellence. The immediate issues focus on the reader: Are you innumerate, even a little? If so, you might start by considering the strange proposition advanced in Chapter 11 that you are already a mathematician. If so, what better way to combat math anxiety? The last chapter offers immediate remedies in the form of light armor against the most common numerical slings and arrows. It analyzes many of the examples and explains certain basic ideas and procedures that must sooner or later inhabit every numerate mind.

Hundreds of people have sent me examples of abuse. I call them math abuse detectives or just abuse detectives. In the examples that appear in the following pages, I have used the names, dates, and places supplied by the detectives themselves. This policy has the advantage of giving the examples immediacy, serving notice on abusers that their practices will be coming under increasing public scrutiny. The policy has one disadvantage. It may lead readers to assume that the names represent isolated cases of abuse when, in fact, the opposite is true. The abusers mentioned in this book are not extraordinary, but typical. They stand in proxy for themselves and their colleagues. As every abuse detective knows, you can look at almost any newspaper or magazine, any ad, any political debate, any corporate or government campaign of public persuasion to find math abuse.

As Shakespeare says, ". . . 'tis the sport to have the engineer hoist with his own petar." There are those who notice mistakes in passing and those who buy books for the sole purpose of finding them. I hope we have eliminated all the errors. Ironically, some deliberate oversimplifications will remain. In writing about mathematics for a large audience, you must maintain the precarious balance between simplicity and truth, almost to the point of becoming an abuser yourself!

A. K. Dewdney

Contents

Introduction

Mathematics lies at the very heart of our scientific and technical civilization. The ability to make and use mathematics is the single, common, indispensable ingredient to all the sciences and every branch of technology. I am hardly the first mathematician to point out how strange it is that such an important subject gets so little support in a part of the world that stresses science and technology.

It is easy to understand why people don't rush out to buy mathematical journals, poring over the latest results in the cohomology of fiber bundles. After all, mathematics does get pretty abstract and seemingly far removed from life. But it is difficult to understand why so many people must struggle with concepts that are actually simpler than most of the ideas they deal with every day. It is far easier to calculate a percentage than it is to drive a car. The notion of probability is child's play compared to the concept of a bridal shower.

Thanks to authors like Douglas Hofstadter and John Allen Paulos (among many others), innumeracy is now widely accepted as the mathematical parallel of illiteracy. Innumeracy, the inability or unwillingness to understand basic mathematical ideas involving numbers or logic as they apply in everyday life, imposes a grave handicap on its victims. In some ways, the handicap is far graver than illiteracy.

Innumeracy has become a major problem in North American society. As our educational system continues its steady decline, as our population slides slowly into the "unthink" of pop culture and instant gratification, people are losing on every front. Their innumeracy is costing money, property, and freedom of choice. Even lives sometimes hang in the balance.

Out there, in the streets that we travel, in the offices where we work, and in the newspapers we read and the television programs we watch await forces that would mislead and misinform us, exploiting the enormous public confusion over subjects like percentages, averages, fractions, compounding, and other basic mathematical ideas. This situation has no parallel in illiteracy. Literacy, after all, concerns a translation skill—learning to move easily between written and spoken speech. Numeracy concerns thought itself. You might exploit people's innumeracy through an advertisement, for example, making a claim that seems to be valid but isn't. But how would you exploit their illiteracy through an ad they can't even read?

Individuals, companies, special-interest groups, the media, and even governments abuse mathematics to sell products and propositions. Math abuse, as I call it, exploits innumeracy by twisting logic and distorting numbers. As this book will demonstrate, it isn't always clear whether abusers of mathematics really understand what they're up to. If they did, we would have a right to be angry. But, sometimes despair seems the more appropriate reaction.

It will take two chapters just to lay out the major forms of math abuse in today's world. Chapter 1, for example, explores how numbers, counting, percentages, and other simple mathematical ideas are twisted to suit private agendas. The most abused branch of mathematics, statistics, needs a chapter all to itself—Chapter 2. Chapters 3 through 9 focus on various areas of abuse such as advertising, lotteries and gambling, finance, business, government, health care, and the media.

Two final chapters attempt to remedy the abuses by laying out a brief course in self-defense for readers who would like to be mathematically streetwise. There are logical meditations and numerical kung fu.

The book's title, *200% of Nothing,* barely hints at the range of abuses. It's time now to buckle your mental seat belt in preparation for a sometimes shocking, sometimes amusing ride through the streets of math abuse. Here are nothings dressed up as somethings, somethings dressed up as nothings, and everything in between.

1

Innumeracy and Math Abuse

Light Bulbs that Generate Power

The magazine ad was both catchy and impressive. It began with the headline:

> HOW MANY LIGHT BULBS DOES IT TAKE TO CHANGE AN ELECTRIC BILL?

Sponsored by Northeast Utilities, a group that includes the Connecticut Light and Power Company, the Western Massachusetts Electric Company, and three other utilities, the ad promised wonderful savings with its new, high-tech, energy-efficient light bulbs and fixtures.

For the purchase of each bulb or fixture, Northeast Utilities offered a rebate ranging from $4 to $50. The offer was solid, money in the bank. But the ad also promised extraordinary savings "on energy," and this is where the problem lay. One of the screw-in fixtures, for example, would save the energy-conscious consumer "65 percent on energy." As if to anticipate the alert consumer asking, "Savings compared to what?" Northeast Utilities placed a footnote at the bottom of the ad:

> *Compared to incandescent (standard) lighting.

The company appeared to claim that if a standard light bulb used 100 watts, the screw-in fluorescent bulb would require 65 percent less, namely, 35 watts. This certainly seemed like an attractive savings, but it was nothing compared to what the ad promised a little further down. There, by installing a metal halide fixture the consumer is offered a $50 rebate *and* promised an amazing savings of "200 percent on energy."

This is a truly astonishing claim and, before falling prey to the wonders of modern advertising mathematics, we are well advised to pause and reflect on the logic involved. To save a mere 100 percent on energy, after all, means to save *all* of it. A light bulb that saved 100 percent on energy would burn brightly without absorbing the slightest amount of power, a physical impossibility. But the ad claimed something even more preposterous. Were the metal halide bulb to save 200 percent on energy, not only would it consume no energy, but it would actually produce an excess 100 percent of energy!

The exaggerated claims by Northeast Utilities were brought to my attention by a *math detective,* a term I apply to the people who have sent me examples of math abuse from all over North America and around the world. Here is an example of what the best sort of math detective does when he or she discovers an abuse of mathematics—and people.

The math detective became upset enough by the ad to telephone Northeast Utilities and ask whether the company would be paying him every month that he used the fixtures. After all, his new high-tech bulbs would be pumping energy back into the power system. "No," the person on the telephone said. "We will not be paying you."

The detective knew the correct way to calculate savings as a percentage. You take the amount saved, say 65 watts, and divide it by the amount previously used, say 100 watts. The resulting fraction, or ratio, must be a number between 0 and 1, in this case, .65. The corresponding percentage, in this case 65 percent, must lie between 0 and 100 percent. In any case, you cannot save more than you previously used.

With the true grit that would make math abuse a thing of the past if we all had it, the detective doggedly telephoned the advertising agency to see whether there had been a mistake in the ad. "No mistake," they told him. He tried to explain what a

200 percent savings meant, but the agency representative had it on "good authority" that the mathematics was all "correct."

In this case of abuse, however, there is no question of correct math and faulty interpretation. The math is clearly wrong. When math is mangled in this way to enhance its persuasion power, we can easily enough discover *that* something went wrong, but figuring out *how* it went wrong can be more difficult. Can anyone tell, just by looking at a car that won't run, what is wrong with it?

Peering under the hood of an abuse, though sometimes perplexing, can make an enjoyable speculative game. An armchair abuse detective, for example, might quickly discover if there is any method in the madness of a "350 percent savings." How did the people responsible for the utilities ad arrive at their savings figures? Did they simply exaggerate the numbers or did they discover a new way to pervert percentages?

Suppose for a moment that the geniuses behind the ad stumbled on a way to calculate percentages that produces more impressive figures than the standard approach, what I will call a *percentage-pumping formula.* Here is one possibility: Suppose they divided the amount of energy saved not by the amount previously used but by the amount unsaved, so to speak. In other words, they divided by the amount of energy the bulb actually required to operate. How would this work? Look at the case of the metal halide fixture. If the fraction of energy saved by the bulb equaled two-thirds, say, then the fraction unsaved would equal one-third. The percentage-pumping formula divides the saved fraction (two-thirds) by the unsaved fraction (one-third) to get 2. Multiplied by 100 to get a percentage, 2 becomes 200 percent.

I hesitate to publicize the percentage-pumping formula. What unscrupulous practitioners of public persuasion might get hold of it? Percentage-pumping can be applied to almost any situation that involves ratios. As far as calling such math "correct," as the agency representative did, we may as well call murders "well executed."

This excursion into percentages illustrates some of the typical features of a math abuse. The best way to outline the subject as a whole, however, is to start with numbers and counting, watching as math abusers overcount to inflate numbers, then undercount to avoid embarrassing information. The subject moves on to ratios, fractions, and percentages. Here the abusers add and subtract percentages with abandon to produce numbers that look impressive but are wildly wrong. The next area concerns the compound growth of numbers. The innumeracy I call *compound blindness* encourages some abusers to impress you with growth that is actually normal. When numbers get really big (or really small), number numbness sets in and math abusers stand ready to make exaggerated claims, to make even more than 200 percent out of nothing.

As the following example shows, you can't always count on the media to count correctly.

Death by Aftermath

It was September 10, 1965. Hurricane Betsy had just ravaged the Gulf Coast, including the city of New Orleans. When power was finally restored, thousands of frightened people watched reports on the storm's aftermath on local television stations. As our abuse detective noted in his files, one newscaster announced a coroner's fatality count of 50 for the New Orleans area. The same reporter allegedly went on to say, "But our studio counts 93." Getting the jump on coroners is no mean feat, as they are noted for the thoroughness and care with which they gather their facts and observations. How did the announcer do it?

Simple, according to our detective. The studio apparently called each of the agencies involved in the emergency—the police, fire department, Coast Guard, and so on. The caller obtained the number of bodies encountered or collected by each agency, then added these to the number of bodies at the morgue. Since some bodies had been reported by one agency, collected by another for sending to the morgue, and then counted at the morgue, it was hardly surprising that the announcer's body count was nearly double what it should have been. The station had probably added the numbers correctly but used flawed logic in gathering them. The station had succeeded in counting two or more deaths for most of the hurricane victims.

When you add together two or more numbers from the real world, you must be careful about what they mean. For example, when adding the numbers from different populations or sets, you must make sure that the populations don't overlap.

As this example dramatically demonstrates, abuse is not always intentional, nor is numeracy on the part of the public always a defense against it. What reason would people have to disbelieve the announcer's body count and how could they know better?

Death by Filtering

Another common distortion involving numbers is to undercount or even to omit numbers. Two examples illustrate the practice—one merely amusing, one deadly.

We perform calculations every day, far more than we realize. Every conclusion we reach, every decision we make, has a calculation of some sort behind it—whether we are conscious of it or not. John Allen Paulos, author of the well-known book *Innumeracy,* coined the term *filtering* to describe what happens when we omit a crucial number from everyday calculations.

One such calculation was performed by a friend of mine, a nurse, when she looked out of the maternity ward window one night and saw the full moon. An unusually large number of babies had arrived during her shift. The full moon merely confirmed what she and a lot of other people believed: More babies are born during the full moon than at any other time.

A miniature controversy currently rages on this subject. Some investigators claim to have found a slight correlation between birth rates and the lunar cycle. Others doubt it. Personally, I find it a perfectly charming idea and have nothing against the notion. But the effect, if it exists, is far too slight for a casual observer to detect. How then, did the nurse "know" that more babies were born during the full moon? She had probably filtered out the numbers born when the moon wasn't full. For example, she may have glanced out the window during the last run of babies, seen the crescent moon, and then forgotten all about it.

Popular beliefs and superstitions frequently find support from just this kind of selective counting. We typically remember only the cases that support our belief while omitting, forgetting, or not even noticing the cases that undermine it. The situation becomes serious, however, when the beliefs concern more crucial issues such as personal health. Things we do not filter for ourselves, others are only too eager to filter for us.

A number of cancer clinics in Tijuana, Mexico, offer cures based on questionable medical practices, for example, the infamous coffee enema treatment. They have pursued business for many years, and continue to operate today, by employing what amount to filters that lure prospective patients to them. The brochure of one such clinic advertises that "50 incurables" came to them with inoperable cancer but were subsequently "cured." The brochure does not tell prospective patients how many people were treated unsuccessfully. Typically, neither does the clinic.

According to William Jarvis, president of the National Council Against Health Fraud (NCHF), there have been enough patients who were not cured by the clinic to support a charge of filtering against the clinic. As he puts it, "Quacks often deceive themselves by using selective affirmation." The NCHF, a worldwide organization of health care professionals and consumers, has tracked dozens of questionable treatments for a great variety of diseases including cancer, a major menace in North America today. Hundreds of times a day, all across the continent, people with serious cancers seek alternative therapies for their disease. Impatient with standard medicine, they may consider alternative therapies and end by throwing logic to the winds.

Such therapies may include untested drugs, rigorous and unusual diets, even psychospiritual methods. In the guarded language of science, the best that can be said for these methods is that no careful, controlled study has been able to confirm the efficacy of any of them. Meanwhile, anxious seekers flock to alternative clinics as their last hope, often foregoing palliative care in the process. The United States Food and Drug Administration estimates that up to $3 billion is spent annually on ineffective cancer treatments. By contrast, the U.S. National Cancer Institute enjoys a budget of less than $1 billion.

Are the unfortunate patients innumerate? A recent study upset researchers' expectations when it revealed that patients undergoing alternative therapies are better educated, on average, than patients under conventional care! Education, which comes under scrutiny at the end of this book, clearly failed these patients if its purpose was to make them numerate enough to spot a filter.

If the manipulation of single numbers creates problems for so many of us, how about double numbers? A ratio or fraction always involves two numbers, one called the numerator, the other called the denominator.

$$\frac{\text{numerator}}{\text{denominator}}$$

A ratio may appear in this form, as a single number, or as a percentage. In all three cases, it implies the comparison of one number with another and therefore cannot always be treated as a single number. Ratios and percentages have an intimate but simple relationship. You can always turn one into the other by either multiplying or dividing by 100. Most people find percentages to be more meaningful than ratios, but they may still suffer from *ratiocinitis,* a tendency to forget about a ratio's double character in adding or subtracting ratios as if they were simple numbers. Not surprisingly, most of the cases of ratiocinitis in this book involve percentages.

Instant Wealth and Rebounding Grades

A man by the name of Smith was walking home from work when he spotted a $5 bill on the pavement. He looked around, picked it up, and put it in his pocket. His other pocket already contained a $10 bill. Smith smiled. "My wealth has increased by 50 percent," he said to himself.

Unfortunately, the pocket that held the $5 bill had a hole in it. When Smith got home, he discovered to his dismay that the $5 was missing. "That's not so bad," he said. "Earlier, my wealth increased by 50 percent, now it has decreased by only 33 percent. I'm still ahead by 17 percent!"

This funny story was reflected in a letter from a California abuse detective who finds that local, state, and other educational authorities are not above abusing percentages in the following manner: "Consider the claim: although in the seventies test scores were down nearly 60 percent, they have since rebounded by over 70 percent.

Such a claim succeeds, like the fictional Mr. Smith, in making something out of nothing. If you listen to the numbers and don't think about them too much, it sounds like the state is doing a better job than ever. Aren't scores 10 percent higher now than they were in the 1970s? No more than Mr. Smith gained any money by first finding, then losing, the same amount.

In order to understand this kind of error, take a minute to perform a simple experiment. Make up a test score, say, 80. In order to calculate what the score would be if it decreased by 60 percent, you must calculate 60 percent of 80 (namely, 48) and then subtract it from 80. The result is 32. Now, what happens to this miserable score when it "rebounds" by 70 percent? It increases by 22.4 (70 percent of 32). The new score is barely 55, a long way from the 80 you started with. In fact, it is only about 68 percent of 80.

Some people pile percentages, adding or subtracting them indiscriminately to obtain favorable numbers. A time series of percentage-changes in a number can never be simply added up to get an idea of "where we are now." After all, each percentage change is based on a new number in the sequence.

Sometimes, it does make sense to add percentages. For example, if you want to find out the percentage of automobile accidents caused by substance abuse, you *may* add up the percentages of accidents due to drinking, those due to cocaine use, the ones resulting from LSD, and so on, as long as the different percentages count completely separate parts of the same population—in this case, automobile accidents.

Sometimes, you *must* add up percentages. It might help to detect the abuse called *number inflation*. The media all too happily pass along questionable figures from public agencies, which we would like to regard as reliable. Number inflation (or deflation, as the case may be) is the most common abuse from these sources. Any number—a percentage, a dollar figure, a head count—can be inflated. And whether we believe the number or not, we often have no way of checking it directly. We are at the mercy of the source.

Drunk, Drugged, Depressed, and Dangerous!

One of our keenest abuse detectives conducted an informal survey of traffic fatality statistics gleaned from various newspapers. The survey revealed the following startling numbers: Depending on which federal or state agencies are consulted, 20 percent of fatal traffic accidents are caused by cocaine, 25 percent are caused by marijuana, 50 percent by alcohol, 35 percent by sleepiness, 85 percent by speeding, 50 percent by smoking, 35 percent by suicide, and 20 percent by mechanical failure. The contributors of these percentages include the Drug Enforcement Administration, the California Highway Patrol, the National Highway Traffic Safety Administration, and the National Safety Council.

How are such percentages possible? If the various categories of cause did not overlap at all, one would expect the percentages to add up to no more than 100 percent. But these percentages add up to 320 percent. This number makes sense only if the great majority of the accidents had several separate "causes," each of them a major contributor to the fatality.

The typical accident might well result from the following astonishing combination of causes: the driver first takes some cocaine and marijuana, then drinks a bottle of wine, then heads out to the freeway in a mechanically unsound vehicle, lights up a cigarette, drives at high speed, feels a suicidal urge, and aims for a bridge abutment, falling asleep just before impact. A simpler explanation might be that the agencies reporting these figures were exaggerating. Why? Agencies compete for funding by making their individual mandates seem as important as possible. Such number inflation is potentially one of the worst math abuses since not even the numerate can guard against it.

The Irrelevant Fund

Numbers do not always grow through exaggeration. They also grow by the perfectly natural process called *compounding*. The process can sneak up on you if you suffer from compound blindness.

If you start with a number and periodically increase it by the same proportion, it can grow, in time, to astronomical proportions. Unless you have some familiarity with the process, you can be taken for a ride when overzealous financial advertising shows how every $1,000 you invest today swells to $10,000 by the time you retire. It is a remarkable growth, but it certainly doesn't result from any special financial acumen on the part of the company that would take over your nest egg.

Compound blindness creates numerous other opportunities for financial scams. Here is just one of them. A double abuse, it hides the compound growth behind a smoke screen put up by a car salesman.

Two abuse detectives have reported this nasty trick, and I have heard of it from a variety of other sources. I've been told of two specific instances, in Florida and Ohio, and I would guess that most automakers have been guilty of the practice in their dealerships at one time or another. Here is how it works.

One of the detectives entered a dealership sales office ready to write a check for $10,000 of his carefully saved money. The model he wanted was out in the lot, waiting. The salesman's eyebrows shot up when the customer offered to pay cash for the car. "Why do you want to pay cash when you could finance the car through us?"

The detective replied that he wanted to save the cost of financing. Even if he were to finance the car, he said, he would do it through the bank anyway, since the bank only charged 7.5 percent instead of the dealership finance rate of 11 percent. As the detective tells the story, the salesman didn't even blink. "But don't you realize that you could still save money by financing with us?"

The detective was not prepared for this. He thought that the dealership would be grateful to receive a cash payment. He was further dismayed to see the salesman swing over to a computer terminal and type in some numbers. The computer paused for a moment, then regurgitated the numbers in the form of intimidating financial data.

The printout showed that a monthly payment of $327 would pay off the $10,000 car in 36 months with a total financing cost of $1,786. The statement further claimed that the customer's $10,000, left in a savings account at 7.5 percent, would earn $2,514 over the same period.

"Look at that! If you leave your hard-earned money in a term deposit, you can make the difference between $2,514 and $1,786. Let's see. That works out to $728."

Thinking furiously, the detective hesitated while the salesman gave him a look reserved for small children and fools. "But, uh, what about the $327 that I pay every month? Where's that going to come from?"

The salesman spread his hands and laughed heartily. "Look, it's not my business where the money comes from. Frankly, you're the first person I've talked to who doesn't understand the system. But, hey, it's your decision!"

The abuse detective recovered his composure, certain that he had found the chink in the salesman's armor. They argued but the salesman did not relent. The detective passed up the purchase and went home with the printout in his hand. At home he got out his trusty calculator and went to work. Suppose he bought the car with the $10,000 outright, but built a fund with the monthly payments of $327 the salesman had wanted him to make on the car. He added in payment after payment, compounding at the bank's rate of interest as he went. In about ten minutes he had the answer. After 36 months the fund would have swelled to $3,171.

Next day, he returned to the dealership to present his results to the car salesman. By buying the car outright and using the monthly payments, he would be $3,171 richer, much more than the mere $728 the salesman had claimed for financing through the payments and investing the $10,000. By neglecting to mention this, the salesman had tried, in effect, to bilk the detective out of the difference in the two amounts, namely $2,443.

Now it was the salesman's turn not to understand the argument. He kept claiming that the fund was irrelevant to the debate! To this day, the detective isn't absolutely certain whether the salesman was innumerate or stubborn.

A Million Drops in the Bucket

When numbers get really big, whether they grow by compounding or just suddenly leap out at us from newspaper headlines, we may be victimized by abuses related to the sheer size of the numbers.

Douglas Hofstadter, originator of the term *innumeracy*, also coined the phrase *number numbness* to describe a widespread inability to appreciate numbers that are either very large or very small. Who, for example, hasn't confused a million with a billion on occasion? After all, as words, they differ by only one letter.

Politicians love to accuse their opponents of misspending "millions of dollars." "My opponent, while in office, misspent the hard-earned money of taxpayers on programs of dubious value. Spending like a drunken sailor, he ran up a debt of no less than three million dollars!" The effort at number bludgeoning almost always pays off because to most people a million dollars sounds nearly as large as a billion. Intellectually, they may know that a million is only one-thousandth of a billion, but they have trouble feeling the enormity of the difference. To *feel* the difference between the two numbers means that numbness has been overcome.

One way to appreciate large numbers involves a simple number line. If the length of the line below represents a billion dollars, the invisible and near-microscopic portion of it on the left (between the ticks) represents a million.

We are equally numb to small numbers, like millionths and billionths. John Sununu, former White House chief of staff in the Bush administration, was known for his sharply critical attitudes on the subject of innumeracy in Congress. In Sununu's view, policy makers "all too often do not have an intuitive sense of the difference between a million and a billion. . . . It is not a trivial change of one letter to write a regulation in terms of parts per billion instead of parts per million." The difference can be seen in the number line above. If the whole line represents one millionth of something, the tiny piece at the end represents a billionth. Suddenly you realize that the difference between a millionth and a billionth is just as crucial as the difference between a million and a billion.

When it comes to really big numbers, there is the *googol*, specially invented decades ago by George Gamow, the physicist and science writer, for his daughter who wanted to know a *really* big number. A *googol* is defined as one followed by 100 zeros:

1000000000000000000000000000000000
0000000000000000000000000000000000
000000000000000000000000000000000

You can write a googol by using a special mathematical shorthand called the *power notation:*

$$\text{one googol} = 10^{100}$$

Ten raised to the 100th power, after all, is just one followed by a hundred zeros. Most scientists, engineers, and other technically minded people have heard of the googol. Didn't they all read Gamow at one time or another?

The National Security Googol

The popularity of the googol may have prompted the National Security Agency to use it in an ad designed to attract computer scientists to its ranks. The ad, which appeared in a number of engineering and science specialty publications, proudly displayed the googol written out across the page. "We're showing you a googol for one simple reason," said the ad. "To consider the National Security Agency, you need to think big."

How big was that? "Counting 24 hours a day," said the ad, "you would need 120 years to reach a googol."

It struck one abuse detective who spotted the ad that the NSA had somehow underestimated the size of the googol. He asked himself how fast he would have to "count" to reach a googol in 120 years. It turned out he would have to count up to an astronomical number *every second*. How many people can count to 2640000000000000000000000000000000000000000000000
00000000000000000000000000000000000000000000000 every

second? He decided not to work for the NSA after all. You not only have to think too big, you have to work too fast.

Obviously, somebody at the NSA (or was it an ad agency?) could not deal with really large numbers. That somebody didn't understand the power notation.

The average person can come to understand and use the power notation in only a few minutes. If everyone understood it, we would all have a much better grasp of large numbers. We would all understand that an increase of one in the exponent (the number in the superscript position) means a tenfold increase in the number itself. The Richter scale that measures the force of earthquakes is based on this relationship. The number reported for an earthquake reflects the exponent of its seismographic motion when written in the power notation. An increase of one in the Richter force means a tenfold increase in the motion of the quake as registered on a seismograph. The force 7.1 quake that shook the San Francisco Bay area in 1989 was a midget compared to the quake that hit in 1906. Estimated to have greater than force 8, the latter had ten times the seismographic motion—which works out to 31 times the power!

We are always impressed when we hear a number that is much larger than we think it should be. Compounding might impress us with the enormous numbers it can produce. But simple multiplication can do the same trick, especially if someone uses it to calculate an area or a volume. People often have a very poor sense of how many square feet their house occupies or how many cubic feet their refrigerator contains.

The Incredible Expanding Toyota

Mention of areas and volumes reminds us of geometry. This is another field ripe for innumeracy and math abuse. We can be influenced by the numbers that areas and volumes produce, or we can be influenced directly by geometry itself. The first abuse concerns volume.

In this case, we encounter the innumeracy I call *dimensional dementia*. Understanding how areas and volumes behave can be tricky, which creates fertile ground for those who would impress or confuse us. Here, for example, is how a recent television ad for the Toyota Camry began:

> How can it be that an automobile that's a mere nine inches larger on the outside gives you over two feet more room on the inside? Maybe it's the new math!

By appealing to "the new math," an experimental method for teaching mathematics introduced in the 1970s, the advertiser sent a signal to the innumerate public: "Don't even try to think about this one. It's beyond you!"

The new math, an educational fad that is happily fading, promised to enhance the scientific and mathematical literacy of a new generation by introducing the language of set theory, logic, and other mildly arcane topics early in the curriculum. Meanwhile, the old math, specifically the simple geometry of volumes, handles the Toyota miracle pretty well.

A 9-inch increase in any of the Toyota's three principle exterior measurements, (height, length, width) cannot possibly produce a change of 2 linear feet in any one of the interior dimensions. This suggests that the "two feet more room" refers not to length but to volume, two cubic feet to be precise. So should we be impressed when a mere 9-inch increase on the outside produces an additional two cubic feet inside? In fact, we should expect an even larger increase.

Look at the dimensions of the generic passenger compartment shown in Figure 1, which are 3 feet (height(by 6 feet (length) by 4 feet (width). The volume is: 3 x 6 x 4 = 72 cubic feet. Any way you calculate the volume with an added 9 inches, the resulting increase is much larger than 2 cubic feet. Increasing its width by 9 inches, for example, increases its volume by a whopping 13.5 cubic feet: 3 x 6 x 4.75 = 85.5. Why, then, did Toyota claim such a small increase in the volume of its compartment? Did the company perhaps simply get its figures wrong? Must be the new math!

Chart Abuse

Numbers lurk in all figures, geometric or otherwise. Maps, for example, imply distances. If you know the scale of a map, you can get a rough idea of how far apart places on the map are. The same thing is true, in aces, of the charts produced by companies, the media, governments, and just about anybody else who

Figure 1 The Incredible Expanding Automobile

would impress us visually with numbers. Such representations, as pointed out by Edward R. Tufte in his classic work *The Visual Display of Quantitative Information,* offer new scope to math abusers. In figurative form, the numbers are now once removed from cold, hard digits. To distort the numbers, chart abusers only have to distort the geometry.

Examples abound of stock market charts that seem headed for infinity and CO_2 readings that make you want to run for the hills, all through the creative application of geometrical distortions and other visual tricks designed to mislead chart readers.

Most people have seen chart abuse in magazines and newspapers without necessarily being aware of it. A good chart fulfills certain minimum requirements, such as having both the horizontal and vertical axes or reference lines labeled with meaningful numbers. You have every right to be suspicious when one of the axes, particularly the vertical one, is not labeled. An abuse detective passed on the wonderful example of chart abuse he found in a corporate annual report, shown in Figure 2. There are plenty of examples of chart abuse, but this one amounts to something of a classic.

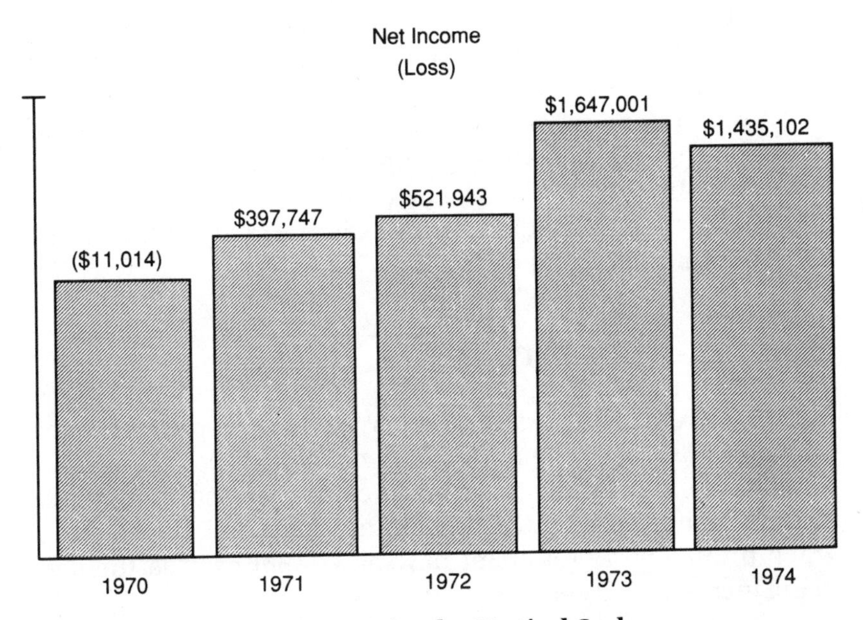

Figure 2 Annual Report Omits the Vertical Scale

In bar charts, the vertical columns or bars each represent a number. The lengths of the bars are normally proportional to the numbers so that you can see at a glance how things stand. At first glance, the corporate earnings seem to be well up. True, there was a slight drop in the current year's earnings but, on the whole, the company looks very healthy.

It looks healthy until you notice the parenthetical earnings of $11,014 above the first vertical bar. In accounting practice, parentheses mean negative numbers. In other words, the company suffered a loss of $11,014 in 1970. How is this reflected in the chart? It isn't. The axes are unlabeled and it takes only a little detective work to see why.

On the basis of the given figures and their bars, you can deduce that the bottom of the vertical axis should receive a label of about *minus* $4,200,000. Even the least numerate of shareholders would have scratched their heads at that one. And numerate readers would have rocketed straight out of their chairs, ceilingward.

But shareholders want to believe their company is doing well. If you don't confront them too directly with the evidence of numbfiscation, they may not even blink. They want to believe what the upward march of those mighty bars seems to tell them: All is well.

Visually, it's a complete lie. In general, we all have the right to expect that bar heights will be proportional to the numbers of interest. In annual reports, the bottom line must be zero, so to speak. If a company has negative earnings (a loss) they must use a bar that hangs below the line.

Graphic Distortion

Many clever visual tricks can be played with graphs either for dramatic effect or to put the best possible spin on negative numbers. Some of these tricks involve outright distortion, while others are more subtle. One disturbing example is the case of a graph printed in a prominent newspaper that charted the dramatic increase in the number of cases of malignant melanoma (skin cancer) between 1935 and 1984. The graph was made more dramatic by the use of a clever illustration that depicted a skeleton lying in a lawn chair, as if out for a day at the beach. The edge of the lawn chair formed the curve of the graph, and the skeleton leered menacingly out at the reader. The depiction was certainly clever, and it must have very effectively drawn the readers' attention, but the result of the dramatization was to create a distorted impression of the actual rate of growth of cancer. In order to accommodate the figures of the skeleton and lawn chair, the graph was stretched vertically, which made the rate of growth look more dramatic than it was.

A similar but more subtle distortion can be seen in Figure 3, where visual tricks are used to make very mildly positive numbers look much better. The figure shows a purely hypothetical advertisement by the North American Winter Resort Confederation, a purely hypothetical organization. A mild upturn in bookings over the last two years is made to look like a rush for the peaks. The ski jump is shown in perspective so that the jumping ramp at the end dominates the picture. The booking numbers that label the vertical line at the left of the chart show an industry in trouble, but the visual impression created by the relatively enormous jump zone (not to mention the skier) is very positive.

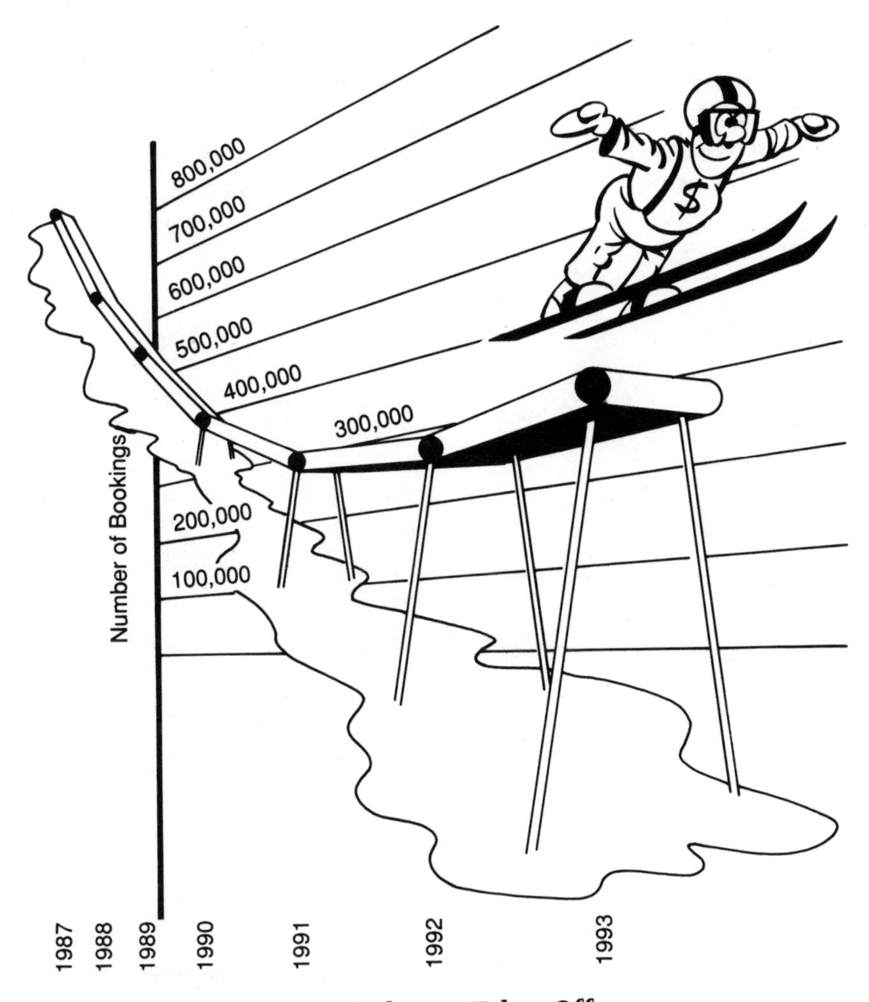

Figure 3 Winter Resort Industry Takes Off

Distortions and visual effects of this kind are not uncommon in the innumerable charts and graphs used to impress us by all kinds of organizations, from private corporations, to the government (as we'll see in Chapter 7), to well-intentioned nonprofit public service institutes. Our best defense is to cast a careful eye at every clever graph we see.

In this chapter, we have traveled from percentage perversion to chart cheating. Each example stands not only for itself but for many closely related abuses of basic mathematics: numbers, counting, ratios, compounding, enormous magnitudes and tiny ones, areas, volumes, and geometry. In the next chapter we will pursue a side of mathematics not mentioned in this chapter: statistics. In the remaining chapters, the focus will shift from how mathematics is abused to the departments of life in which such abuses are found. In those chapters, you will find close cousins of the abuse types catalogued here and in the next chapter.

Statistics and Damned Lies

Figures often beguile me, particularly when I have the arranging of them myself; in which case the remark attributed to Disraeli would often apply with justice and force: "There are three kinds of lies: lies, damned lies, and statistics."
—Mark Twain, *Autobiography,* Volume I.

Statistical abuse has apparently been with us at least since Mark Twain's day. The phrase, "lies, damned lies, and statistics," has been attributed to no less than five people, including Twain and Benjamin Disraeli, a nineteenth-century British prime minister. Whoever actually coined the term appreciated not only the public's difficulty with statistical ideas, but the enormous potential for statistics to be willfully misused.

Like other abuses of mathematics, statistical subterfuges range from the trivial to the tragic. If a soft drink company twists blind taste tests to its advantage, there are no grave social consequences. But when people are routinely frightened to death by threats from organizations and groups that overstate probabilities connected with health, the environment, or other issues, we should be troubled.

No one will be surprised that many abuses described in this book are committed by institutions and organizations devoted to skimming and scamming the public. Many people may be surprised at the number of worthy and responsible organizations that routinely generate more than their share of chicanery.

My exposé of those institutions and how they and others mislead us will proceed from the simplest manipulations of probability to some arcane distortions of distributions.

Any normal introduction to statistics begins with probability, builds slowly to the theory of distributions, then gets into the science of sampling and hypothesis testing. As a purveyor of street statistics, I will not take you through such a cultivated landscape but through a junkyard of abuses, albeit in textbook order. The first few examples involve probability, the next ones depend on distributions, and the ones after that, sampling. Some of the abuses are mild, some vicious. Some are blatant and some are surprisingly subtle.

The Great Pepsi Challenge

A sampling of damned lies begins mildly enough with a subtle lesson in probability, courtesy of the Pepsi-Cola Company.

Have you ever taken part in a blind taste test? Perhaps you were asked to distinguish two wines, two beers, or, worse yet, two soft drinks. It is difficult to distinguish between Pepsi-Cola and Coca-Cola. I confess to getting them wrong as often as I get them right, and I'm not the only one with this difficulty. The case of the Pepsi Challenge makes an interesting example of the way that probabilities can be deceptive. Did Pepsi really believe that most people would be better at identifying its brand than I am? It didn't need to because probability, all by itself, would do the trick.

Probability, usually expressed as a number on a scale from 0 (impossibility) to 1 (certainty), is a slippery concept. It gives rise to the form of innumeracy I call *norming*, the tendency to interpret probabilities too strictly and not to realize the great range of variation that can underlie a single probability.

The best place to start explaining probability and norming is right in the middle of the scale at the place known variously as "even odds," "50-50," "even money," or "a 50 percent chance." Known to mathematicians as probability 0.5, it has a very simple operational interpretation. An event that has a probabilty of 0.5 will occur (unpredictably) in about half the cases where it has the chance to occur. When you flip a coin, you have a 0.5 probability of getting heads. If you flip the coin a limited number of times, you may not get heads half the time; instead, you may just get *all* heads—or *all* tails. This is what I mean by the range of variation that can lurk behind a single probability like 0.5. People who norm a probability like 0.5 believe that the number of heads will always be half the number of tosses, or very close to that. In fact, the ratio does not generally get close to 0.5 until you have tossed the coin a very large number of times.

Guessing between two alternatives, only one of which can be right, is much like flipping a coin. If you really can't tell the difference between the tastes of two brands of soft drink and someone gives you one of them in an unlabeled glass, your chance of correctly identifying the beverage would be 0.5.

In the late 1980s, the continuing war between North America's two major soft drink manufacturers, Coca-Cola and Pepsi-Cola, took a turn for the worse, statistically speaking. Pepsi-Cola began a campaign of television advertising that attempted to demonstrate decisively the drink's popularity. The audio portion went something like this:

> "In recent side-by-side blind taste tests, nationwide, more people preferred Pepsi over Coca-Cola."

The video portion showed people sampling unlabeled glasses, then smiling in astonishment when the brand they liked best turned out to be Pepsi. No doubt, the commercials had the desired effect on a great many people since "more" sounds impressive in this context. Pepsi may have counted on the fact that most people would not distinguish between what the claim seemed to imply and the relatively minimal "facts" needed to support it. What a television commercial says need only be literally or minimally true to pass muster with those who enforce truth-in-advertising laws, and the truth can be much less impressive than it sounds.

Under the assumption of minimal truth, what do Pepsi's claims really tell us? Start with the first phrase, "In recent side-by-side blind taste tests. . . ." Suppose that more people surveyed actually preferred the taste of Pepsi in just *some* of the tests. These might be the only tests the commercial referred to.

After all, it did not say, "In *all* recent blind taste tests. . . ." This is how the minimality assumption works. If at least some of the tests worked out to Pepsi's advantage, the claim is established. Of course it might well be that Pepsi was referring to all the tests and that they all turned out in Pepsi's favor. The point is, we can't know for sure on the basis of the ad alone. However, this is not the main abuse under discussion at the moment; it is another abuse I call *sample trashing*. It will come up again later in the book.

The minimality criterion, when applied to the word "tests," allows the abuse detective to conclude that there were at least two of them. The analysis continues with the next phrase, "more people." To qualify as "more," what proportion of a crowd should prefer the taste of Pepsi? Frankly, you only need fractionally more than 50 percent.

Keeping these minimal requirements in mind, consider the following scenario. Pepsi ran, let us say, five blind taste tests, each involving seven randomly selected Coke drinkers. Most of these people, quite possibly all of them, could not tell one drink from the other when served cold. They misidentified their favorite drink, Coke, with probability 0.5, that is, about half the time.

We might be justified to compare such blind taste tests to the toss of several coins, one for each person in the test. Each coin has a probabilty of 0.5 that it will come up heads. Adding all coin tosses together, the seven coins will land with a mixture of heads and tails. Here is where the norming comes in. Many people will not stop to visualize the many possible outcomes. Sometimes, for example, there will be more heads than tails and sometimes more tails than heads. Trash the experiments where a majority come up tails, and you can go on television with a commercial of your own: "In recent coin-tossing experiments, more coins preferred heads to tails."

Some people, especially those who believe they would have no trouble distinguishing Coke from Pepsi, might well cry "foul!" After all, I have only *assumed* that no one can taste the difference between drinks. Am I just going to leave it there?

I may as well confess that in the case of more subtle abuses, outside knowledge must sometimes be brought into play. My contention that few people can make such taste distinctions is not entirely a conjecture.

Let's begin with beer. You might suppose that beers would be easier to distinguish by taste alone than soft drinks. After all, beers are made in so many different ways: fire brewed, cold filtered, the old country way and, for all I know, the modern, high-tech way. Moosehead Beer reminds us of the swamps of New Brunswick whereas Coors Silver Bullet apparently begins in a mountain stream somewhere. Naturally, there are those who claim to know all the brews by taste alone, especially after they've been sampling for a while. I wouldn't dream of arguing with them.

A recent informal survey by the Canadian Broadcasting Corporation put the lie to beer-hall braggadocio. A CBC camera crew visited a number of popular pubs and bars and conducted a test using a dozen identical glasses filled with Canadian beers and a few American ones thrown in for good measure. In each watering hole, the crew invited self-styled connoisseurs to identify the brews. The results were devastating. Most of them got most of the brands wrong!

If you tried to draw firm conclusions from such a small sample of beer drinkers, you would be guilty of statistics abuse —the very thing this book is against. But the experiment suggests that most people cannot tell the difference between one beer and another. In fact, the effect is known to science. Combine the relative sameness of taste with the numbing effect of cold on taste buds and you can be pretty sure that almost no one will be able to distinguish one cold beverage from another. Almost everybody, Coke drinker or not, will identify a glass of cold Coke as a glass of cold Pepsi with a probability of 0.5.

Of course, Pepsi may have actually conducted its experiments with perfect rigor, but the truth-in-advertising criteria don't tell us for sure. The point is that nothing more rigorous than the scenario above would necessarily be required to generate the statistics cited by Pepsi. This, in turn, shows us just how meaningful the statistics are.

Lotteries and Lightning

Anyone who has trouble grasping a red-blooded probability of 0.5 wouldn't have a hope of coming to grips with a near-vanishing probability like 0.0000000715. The problem is double-barreled because number numbness also comes into play with such a small value.

Does anyone truly understand how little chance he or she has of winning big in a lottery? Private and state lottery corporations, which annually part North Americans from some 30 billion of their dollars, certainly hope not.

Attempts to explain the fantastically small probability of winning a lottery often meet with the objection, "But what if I'm the one?" It would be fair to respond, "Don't worry, you won't be." To make the point a little more forcefully, one could drag in the old lightning analogy, "Well, what if you get hit by lightning?" Most people would answer, "Fat chance!" But what if you *are* the one to win a lottery? The bait dangles before millions of people who never fail to buy their weekly ticket.

To demonstrate the odds more effectively, let's look at a popular form of lottery that allows the player to select six numbers between 1 and 49, inclusive. The price of a dollar or two seems like a good value if it means a real chance at winning several million dollars. Imagine being a mere six numbers away from a new life. Six little numbers! Alas, the dream is an empty one.

The real chance of winning a six-number lottery, if you buy one ticket, is 1 in 13,983,816. To express this as a probability, take out your calculator and divide 1 by 13,983,816. You will get something like 0.0000000715. This is an extremely small number. How small is it? Number numbness sets in. A mathematician might say that for all practical purposes it is zero. As one wag put it, "You have the same chance of winning whether you play or not."

In case you still don't think this number is hopeless, think about lightning. In the United States, lightning kills between 200 and 300 people a year. The population of the United States is roughly 250,000,000. To calculate the probability of a real event, you must divide the total number of actual cases (say, 250) by the total number of potential cases (250,000,000). With the abandon of professional license, a statistician might declare, "The probability of a randomly selected American being struck and killed by lightning this year is approximately 250 divided by 250,000,000 or 0.000001." One chance in 1 million means that you are a great deal more likely to be killed by lightning sometime in the next year than you are to win the next 6-49 lottery you play. That chance, after all, is about 1 in 14 million.

Death Threats

The probability of death by lightning is much smaller than certain other risks that you take simply by being alive. Until you learn otherwise (by meeting your actual fate), you will be under many probabilistic death sentences, some small, some not so small. If you are a woman, for example, you may recently have been scared to death (well, not quite) by learning of your horrendously high risk of breast cancer.

In January 1991, the American Cancer Society announced that the odds of women getting breast cancer had risen to one in nine. Women all over North America panicked at the one-ninth probability, saying to themselves in effect, "Of course, I would just be the one!" What these women imagined and what the probability really meant were two very different things. As reported in *The New York Times* in March 1992, the Cancer Society had compressed a lifetime probability into a small temporal frame. The Society was guilty of what I call detemporizing, that is, detaching a probability meant for a specific time frame and applying it to another. Women, understandably, imagined that the probability of one-ninth applied to themselves sometime soon, when, in fact, it didn't.

As the story in *The New York Times* made clear, the one-ninth probability applied over an entire lifetime—a 110-year lifetime. The probability of a woman dying, sooner or later, of breast cancer is one-ninth or about 0.11. For a woman under 50, the chance over the next year is more like 1 in 1,000.

The New York Times quoted a Cancer Society spokesperson as saying, "The one-in-nine is meant to be a jolt . . . It's meant to be more of a metaphor than a hard figure." As the article made clear, the "metaphor" is terribly misleading. Dr. Patricia T. Kelly, a risk counselor at Salick Health Care in Berkeley, California, tells her patients that even women in their eighties do not face a one-in-nine chance of developing cancer in the next year. Women under 50 face a risk that is somewhere in the neighborhood of 1 in 1,000 or 0.001 of contracting breast cancer sometime in the next year of their life.

A nasty little scenario I call the grim reaper game illustrates the difference between a probability applied over a short and a long period. Suppose the grim reaper, in an unusually ugly mood, forces you to play Russian Roulette. He hands you a revolver and tells you to spin the six chambers, one of which holds a bullet. When the chambers stop spinning, he takes the gun, holds it to your head, and pulls the trigger with a hollow laugh. The chances are pretty good that the gun won't go off, 5/6 or .83.

Luckily, it doesn't. But the game doesn't end there. The deadly stranger tells you to spin again. And again. Three more times for a total of six, you are told to spin. What is the probability that you will be shot at some point during the six trials?

You can work the number out more easily by first calculating the chance that you *won't* be shot. To get that number, multiply together the six equal probabilities of *not* being shot in each trial, namely 5/6, or 0.83. (I will explain why this is so in the last chapter.) Multiplying them all together, you get 0.335, the probability of *not* being shot in any of the six trials. The probability that the stranger *will* eventually shoot you is therefore the opposite probability, namely 1 – 0.335 or 0.665, a better than even chance. This little example illustrates the relationship between the short-term and long-term effect of a probability. Over time, there is a distinctly greater chance, four times greater in this case, that you will be shot.

The probabilities that you will develop breast cancer in your life, in the next year, or even in the next 24 hours are all quite different. For a given disease, the probability of contracting it over limited periods of time shrinks with the period itself.

What Happens When You Toss Seven Coins 128 Times?

With this all-too-sketchy introduction to probability by way of its abuse, the time has come to introduce distributions, simple diagrams that show you all possible outcomes at once. There is no better antidote to statistical innumeracy.

To appreciate what a probability really means in operational terms, you must know something about the underlying distribution of events, in other words, the complete range of outcomes associated with the probability. It was one thing, in the Pepsi-Cola/coin-tossing experiment, to observe that a coin will come up heads in approximately half the tosses. But it is quite another to understand the variation implied by the word "approximately." When I tossed the seven coins to simulate a group of seven Coke drinkers, I quickly realized that any number of coins from 0 to 7 might come up heads. Often I got three heads or four, but sometimes I got none and once I got six. A distribution diagram lays out all the possibilities at once, enabling you to assess each possible outcome along with the probability of that outcome.

Figure 4 shows what happens when seven coins are tossed many times. You can ignore the diagram on the top, a theoretical distribution, for the moment. The diagram on the bottom shows what actually happened when I threw the seven coins 128 times. The numbers below the vertical bars list the possible outcomes in terms of heads. They range from 0 heads to 7 heads. Above each outcome, the height of the bar gives the number of times I got that many heads. The diagram, called a *histogram,* summarizes the outcomes of all 128 tosses at a glance. The first thing you notice is that it has a hump in the middle, like the distribution on the left, but it is otherwise somewhat ragged in appearance. The hump just means that three or four heads came up more often than the other possibilities.

The theoretical distribution has a beautiful, symmetric appearance. It has the same labels below its bars, but the bars tell you a slightly different story. The height of each bar gives the probability, when you toss seven coins, of getting that number of heads. For example, the probability of getting three heads is 0.2734, a little over 25 percent. Should it surprise anyone that the probability of throwing three heads is the same as the probability of throwing three tails (four heads)? Of course not. That explains the distribution's symmetry. It peaks in the middle and tapers off to either side where the probabilities of getting seven heads or no heads are both about 0.0078.

The real beauty of a theoretical distribution lies in its predictive value. If I toss the seven coins 128 times, the theoretical distribution tells me how many times I should expect each outcome. Just multiply the probability of the event by 128. The

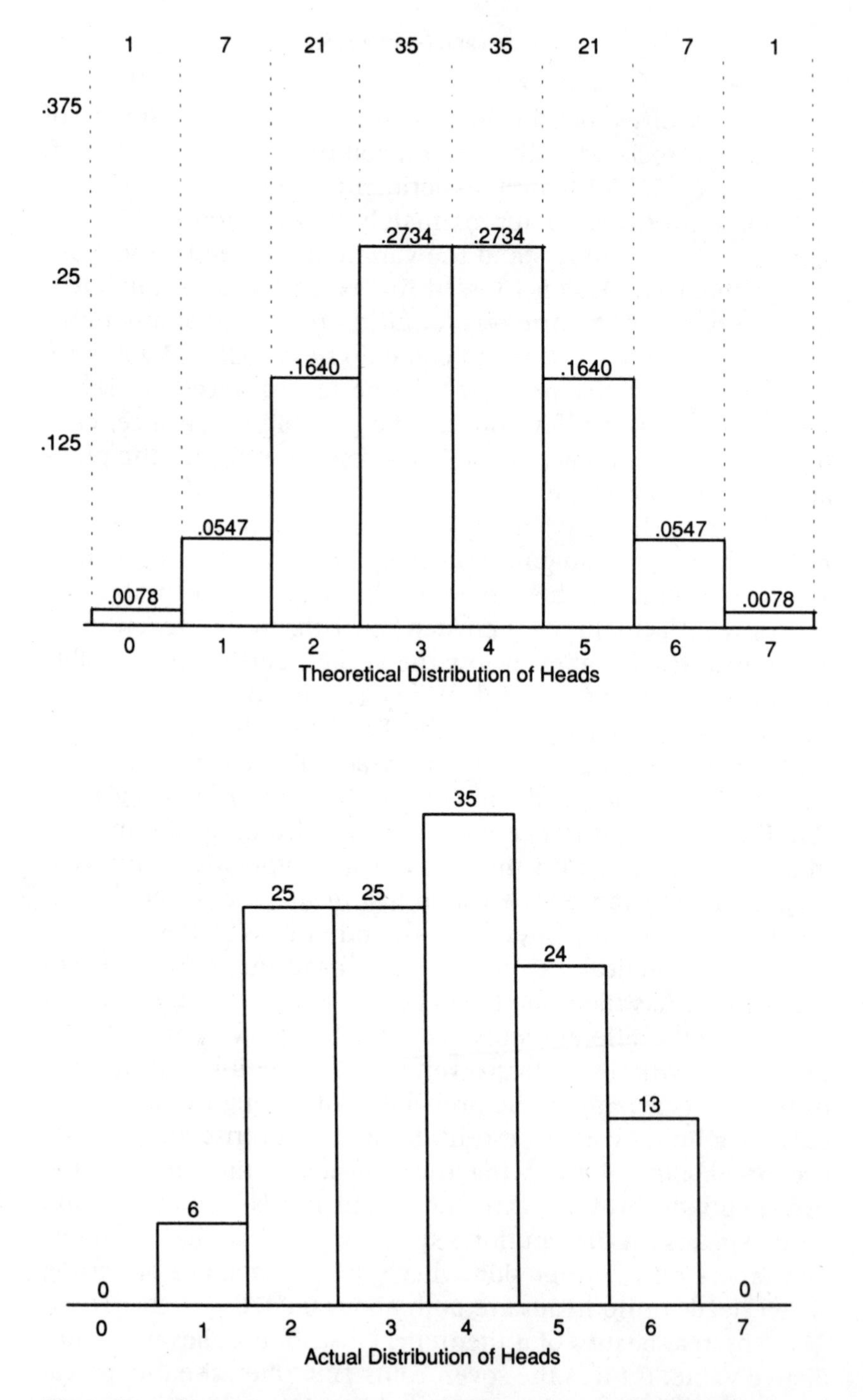

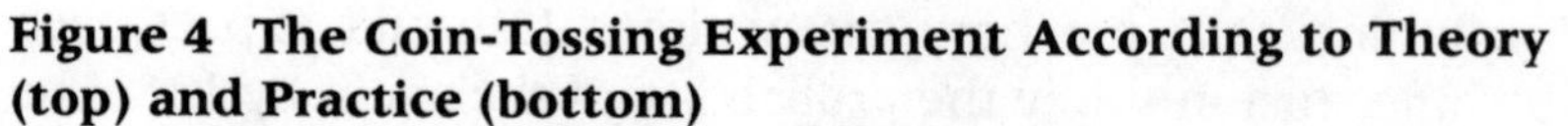

Figure 4 The Coin-Tossing Experiment According to Theory (top) and Practice (bottom)

theoretical distribution predicts 1 case of 0 heads, 7 cases of one head, 21 cases of two heads, and 35 cases of three heads. After that, it makes the same predictions in reverse order, starting with 21 cases of four heads. How close was my empirical experience to the theoretical predictions? I got 0 cases of zero heads, 6 cases of one head, 25 cases of two heads, and 25 cases of three heads (a coincidence). Notice that the numbers I got differed from the theoretical predictions by as little as 1 and by as much as 10, a pretty typical outcome for such an experiment.

There is no better visual summary of the relationship between theory and practice in statistics than a view of these two distributions. The practical or empirical distribution is not symmetrical. It resembles the theoretical one, but only approximately. If I repeated the experiment, another shape would emerge. And if I repeated the experiment again, yet another. All or most of the distributions that resulted from the seven-coin-tossing experiment would resemble the theoretical distribution roughly. In aggregate, they would probably resemble it even more closely.

I urge readers to try the experiment of throwing seven coins 128 times for themselves, but only in the privacy of their own homes. People in public places just aren't used to seeing lone individuals throwing handfulls of pennies into the air, carefully noting the results in a book.

If a company, in a fit of statistical enthusiasm, ran 128 blind taste tests, it might well obtain a result like my coin-tossing experiment. Then the company could really go to town with claims. In fact, it could even zero in on the 13 cases where six out of seven people, about 85 percent, preferred the taste of its product, and it could claim that more than 85 percent of those who drank a competing product preferred the taste of its own product.

The important feature of this distribution for now is its distinctive shape: high in the middle, low on the sides. This means that when you run a multiple experiment such as tossing a number of coins or testing a number of humans, you will often get results that match your expectations. But the outlying parts of the distributions, whether theoretical or empirical, bear the same message: You will also get occasional anomalous-looking results. It is, after all, possible to throw seven heads. And if people really can't tell one cold drink from another, it is also possible to get seven who all prefer the same drink.

The Naked Fund Manager

In recent years, the stock market has boomed, gone bust, and boomed again. Mutual funds have grown enormously in popularity, and many North Americans try to get ahead of the game by choosing a fund that will increase their personal fortune, be it a million dollars or a humble retirement savings plan. Whether they do well or poorly seems to depend on the financial acumen of a mysterious character known as a fund manager. When you turn your money over to a fund, your money might grow by leaps and bounds. Or it might actually shrink. Most people think that what happens depends on the market savvy of the fund manager. Could it be possible that the multibillion-dollar mutual fund industry is based on a statistical illusion?

As most experts on finance are quick to point out, it is relatively easy to make money in a stock market in which most of the prices are rising. It is similarly easy to lose money when prices are falling. Whether a market is rising or falling may be judged by consulting one of its indexes. The Dow Jones Industrial Average, for example, measures performance of the New York Stock Exchange, as does the Standard and Poor 500. Both indexes amount to a weighted average of the prices of certain selected stocks. This means that each stock price is multiplied by a certain number, or weight, before it is added to the average. The Dow Jones incorporates some 30 stocks and the S&P 500 averages 500 stocks, as the name implies. Since it includes more stocks, the S&P 500 makes a better index or proxy for the market as a whole than the DJIA, but a comparison of their performance over time reveals that they frequently move in concert. In what follows, I will pretend that there is just one index, say the S&P 500.

A really "good" fund rises faster than the index; conversely, when the index falls, a good fund may also fall, but not as far. In both cases, because the fund grows more quickly than the index (or decays more slowly), we say that such a fund *outper-*

forms the index. The $64,000 question is whether some fund managers can outperform the index consistently and, if they do, whether you should believe in their market savvy.

The funds that have done well, or outperformed the market several years in a row, do not hesitate to say so in their advertising. They attract a lot of investors and swell to enormous size. Their managers come to resemble financial emperors. Imagine the stir it would create if it turned out that their performances were largely a matter of luck. What would happen if investors everywhere decided that the financial emperors are wearing no clothes?

Strangely enough, it is a well-known fact in the funds "industry" that, in any given year, about half the stock market funds outperform a major index like the S&P 500. This seems to indicate that chance has more to do with a fund's performance than its manager's wizardry. But, you may ask, can probability explain how some funds seem to do well consistently for many years? You might not think so if you normed the probability. You might not realize the extreme forms of behavior it can produce.

Funds that perform consistently well certainly exist. The February 1989 issue of *Money* magazine, for example, carried a revealing assessment of 277 of the best-known, stock-based mutual funds. It tabulated their year-by-year performance since 1979 and found that five funds outperformed the S&P 500 index eight years out of ten. What did the managers of those five funds know that the others didn't?

Just for amusement, imagine 128 funds, each controlled by the toss of seven coins, one for each of seven years. Will some of their "managers" emerge triumphant after seven straight years? Looking back at my histogram (Figure 4), you can readily imagine that each vertical bar logs the number of funds that outperformed the market (scored heads) no years, one year, two years, and so on.

Interpreted as the performance of 128 funds, no less than 13 of the pretend managers outperformed the index six years out of seven! If the 277 funds mentioned by *Money* magazine behaved more or less randomly in relation to the S&P 500, how many would you expect to outperform the index for eight years out of ten? The expected number, worked out in the last chapter, is 12. In other words, by 1989, the 277 mutual funds surveyed in the *Money* magazine report had actually performed slightly worse than you would expect if their managers had simply chosen their stocks at random!

The rage of a thwarted illusion knows no bounds, and, to protect myself, I had better take refuge in the strictest possible truth. Although I strongly suspect (as do other, more knowledgeable writers on financial affairs) that fund performance is ruled more by probability than anything else, I had better not claim that fund mangers who consistently outperform the market are merely lucky. I do *not* know for a fact (nor does anyone else) that there *aren't* mangers who, by employing certain methods, are guaranteed, at least in a statistical sense, to consistently outperform the market. But there can't be very many of them if we get in practice about as many lucky managers as we expect in theory.

Burton Malkiel, author of *A Random Walk Down Wall Street,* claims that there is simply no way to consistently outperform or outguess the market. All price movements are random in the sense of being unpredictable. In science, when you can simulate a real-world result fairly closely with a random experiment like this, you can claim it as evidence of randomness. In other words, if the performance of fund managers cannot be distinguished from the results of pure chance, no one should lie awake nights wondering how the "successful" ones manage it.

A stock market index is what statisticians call a *sample.* It consists of a selection of stocks that many people are happy to accept as a proxy for the market as a whole. Insofar as the index represents the market, it would be called a good sample. Bad samples commonly occur when people practice statistics, and they make fertile ground for a variety of abuses.

Numerical Terrorism at the Nuclear Plant

In 1989, the Toronto *Globe and Mail* published an article that raised concerns about the incidence of childhood leukemia near nuclear reactors. The article made much of a study which found that the incidence of leukemia within 25 kilometers of the Bruce Nuclear Power Station in Ontario was three times the national average. It appeared during a time of controversy, when antinuclear groups were demanding that all nuclear facilities be shut down. Before deciding whether this was an appropriate response to the study, however, we should learn more about it. Conducted by the Atomic Energy Control Board (AECB) of Canada, the study surveyed all nuclear sites in the country and probed the incidence of leukemia in children under age five within 25 kilometers of each site. The author of the article decided not to mention some of the less exciting parts of the report—such as all the other site statistics!

Perhaps it simply wasn't newsworthy to note that the AECB study also found that within 25 kilometers of the Chalk River Nuclear reactors, also in Ontario, the incidence of infant leukemia was only half the national average. Further, the report concluded: "The statistical analysis demonstrated that all of these relative risk values could have occurred due to chance."

By reporting only the frightening statistics and not the others, the newspaper had, in effect, trashed most of the sample. Selective sampling can be made to produce almost any result you want, after all. As far as the study itself is concerned, the sample was already small enough. It did not consist of the relatively large number of people living within 25 kilometers of a nuclear reactor, but of a relative handful of infants with leukemia.

Enthusiastically promoted causes may be supported by enthusiastically distorted mathematics. Some of the abuse that emerges this way may be due to simple innumeracy, but some is probably intended. Some people believe, after all, that if the cause is just, the abuse really doesn't matter. Readers only have to think back to the breast cancer example for a public admission of this belief. In pursuing campaigns of public persuasion, many organizations take advantage of widespread innumeracy on the part of the public.

Hitting the Hites

In matters of public persuasion, no single instrument is used (and misused) as frequently as polls. To obtain an accurate picture of public opinion, you must take a large enough sample to rule out statistical fluctuations, you must gather the sample carefully, and you must ask the right questions. Experience has shown pollsters that missing just one of these criteria invites disaster.

In 1976, feminist Shere Hite published *The Hite Report,* a largely qualitative account of female sexual experience. The book shocked many people with its suggestion that a great many women were unsatisfied with their love lives. The book was followed by two others: *The Hite Report on Male Sexuality* (1982) and *Women and Love* (1987). By the time the third book appeared, Hite found herself in the eye of a hurricane of controversy. Whereas the first book had implied that men were largely incompetent as sexual lovers, the third book seemed to imply that they were distinctly unromantic.

These suggestions surprised and alarmed a great many people, especially men. The subtitle of the third book summed it up: *A Cultural Revolution in the Making.* Was there really a revolution? Was female sexual and romantic dissatisfaction as widespread as the books suggested?

Hite gathered the opinions expressed in her books through two polls. For the first, she mailed out approximately 100,000 questionnaires, mostly to women's groups, women's magazines, and other organizations where she felt there might be women who would likely respond. She got back 3,019 questionnaires and felt justified in generalizing on that basis. For the second book, she mailed a questionnaire to 119,000 men and received 7,000 responses. Not a polling expert by any means, Hite evidently believed that the large number of responses alone justified her conclusions. For example, since 70 percent of her respondents claimed to have extramarital sex, Hite felt free to generalize the percentage to the public at large.

Hite's sampling methodology left much to be desired. David W. Moore, director of the Survey Center at the University of New Hampshire, explains that scientific pollsters consider such low response rates to give almost meaningless results. The number of responses, 7,000, is indeed large but the rate, namely 7,000 divided by 119,000 or 6 percent, is low. In his book *The Super Pollsters,* Moore documents the measurement and manipulation of public opinion. In particular, he shows how low response rates fooled some of the great pollsters of the past. The lower the response rate, the more likely it is that those who respond have some special reason for doing so. In the context of Hite's questionnaire, which invited women to talk about their frustrations, it is probable that women who were relatively happy with their love lives would be far less likely to respond.

A poll on the subject conducted by ABC and *The Washington Post* in October 1987 followed standard polling procedures worked out over a century of statistical evolution. The number of contacts, 1,505, was certainly high enough to guarantee statistically valid results given the response rate of 80 percent. The ABC/*Washington Post* poll contradicted almost every one of the Hite findings.

The public should also be aware that the very questions a pollster asks can have a bearing on the outcome. Among the many examples of statistical abuse and innumeracy compiled by A. J. Jaffe and Albert F. Spirer in their book, *Misused Statistics,* is the following little gem concerning two polls on abortion. In an experimental poll conducted jointly by *The New York Times* and CBS News, respondents were asked two questions on two occasions widely separated in time. The questions were:

1. "Do you think there should be an amendment to the Constitution prohibiting abortions, or shouldn't there be such an amendment?"
2. "Do you believe there should be an amendment to the Constitution protecting the life of the unborn child, or shouldn't there be such an amendment?"

The people in the poll responded differently to the two questions, 29 percent favoring the first amendment and 50 percent favoring the second one.

Occult Samples

People draw many conclusions from samples on a daily basis. If someone cheats you a few times, you might consider it a fair sample of that person's behavior and avoid him or her thereafter. If you find two or three bad apples in the supermarket bin, you might conclude that many more will also be rotten. In both cases, we make a judgment based on the sample as a whole. But sometimes we witness something so profoundly shocking that we forget the rest of the sample altogether.

According to *Reader's Digest,* a man by the name of George D. Bryson once checked into a hotel in Louisville, Kentucky. The desk clerk assigned him room 307. Shortly after settling into his room, Bryson received a letter addressed to Mr. George D. Bryson, Room 307. The puzzling thing about the letter was that he had not told anyone he would be in Louisville. In fact, he had stopped there only on a whim, never having seen the city before.

He opened the envelope and gasped with astonishment. The letter was not for him! Instead it was for Mr. George D. Bryson of Montreal who had recently stayed in that very room!

You can almost hear Rod Serling, host of the famous television show "The Twilight Zone", mumbling against the familiar background theme: "George D. Bryson has just checked into an ordinary Louisville hotel where he will meet his namesake after checking in—to The Twilight Zone."

Of all the abuses of mathematics, of all the abuses of science generally, no single phenomenon causes more misunderstanding than coincidences. They regularly mislead people to the point where they believe in mysterious influences, dreams of the future, and occult powers. For example, a great many people are persuaded that something like telepathy exists because, on occasion, they no sooner think of someone and that very person calls!

All coincidences fall into a broad statistical pattern that originates in extremely large samples. Those who think that coincidences require a supernatural explanation are guilty of *sample occulting*—hiding the size of a very large sample from themselves. The innumeracy was recently exploited by television advertisements for the TIME-LIFE book series called *Mystic Places.* A woman turns to the screen and says something like, "I was just thinking of George and he called!"

Given a large enough collection of things, coincidences are not only not surprising, they *must* happen. Take the case of the mysterious phone call. This phenomenon is just frequent enough for most people to have experienced it. There you are, thinking of X and, within a few minutes, X calls! Isn't that unbelievable?

One of the greatest tools of the numerate are back-of-the-envelope calculations. Sometimes a rough reckoning that uses approximate figures produces answers that, although approximate, are very revealing. Suppose, for example, that you know approximately 200 people well enough to find yourself thinking about each of them occasionally. The 200 people might include family, friends, business associates, and anyone else with whom you might have regular dealings. "I wonder how Phyllis is getting on with her new boyfriend." "Why didn't Mr. Smith deliver those packages?" "Gee, what a good time we had with Ben and Marjorie at that party!" Suppose further that you think about ten of these people a day, and that two of them call you each day, on average. The calls could come at any time.

The phone rings. What is the chance that the call comes from someone you thought of in the previous 24 hours? The answer, according to the formula I introduced at the beginning of this chapter, is the fraction

$$\frac{\text{number of people thought of}}{\text{number of people known}},$$

or 10/200. The probability, 0.05, is small, but not as small as winning a lottery. Now the probability that the call is from someone you thought of in the previous *minute* is much smaller, of course. In the 16-hour (waking) period prior to the call, there are 960 minutes, and the probability is now roughly 1/960 what it was before, namely .0001 or one chance in 10,000.

This might seem to put those mysterious phone calls back on occult status, except for one thing. We are only asking that it happen at least once in, say, a 10-year period. With two calls a day and roughly 10 × 365 days to receive them in, what is the chance that at least one of those calls will be "occult"? As I show in the last chapter, the probability over this period climbs to 0.52, a better than even chance.

The numbers that emerge are admittedly inaccurate because it is only a back-of-the-envelope calculation. But if the assumptions apply even approximately, you can expect to receive a mysterious phone call sometime in your life. I get them about once a year.

This completes the crash course in types of abuse. The next section of the book explores the main areas of life where math abuse occurs, where advertisers, financial institutions, governments, and the media, to name just a few, exploit our innumeracy.

3

The Mathematics of Advertising

Advertising may be described as the science of arresting human intelligence long enough to get money from it.
—Stephen Leacock, *The Garden of Folly*

Everywhere we go we encounter ads. We see them on television, hear them on radio, and glance at them in magazines and newspapers. Every time we encounter an ad, in any medium, we should be on our toes. The person who wrote it may have used a special form of mathematics to sell you goods, services, and even political opinions. I call it *advertising mathematics.*

If the enormous glut of advertising information that surrounds us is any indication, advertising works. By the same token, the prevalence of math abuse in advertising must mean that advertising mathematics also works. Large numbers of people routinely fall prey to faulty mathematics in ads that, on the surface, seem to be spelling out valid reasons for buying. Advertising of this kind not only takes advantage of the public's low level of mathematical proficiency but, at the same time, aids and abets innumeracy through its all-pervasive influence.

Advertisers typically misuse mathematics in three main areas: prices, product descriptions, and quasi-surveys.

The fact that prices are numbers means that they can be minimized, underestimated, or detached from context, as in the claim: "The car is yours for just $199 a month!" Product descriptions often compare numerical measurements such as the size, price, or quality of the product with those of a competing product. In the process, they can make comparisons that imply far more (or far less) than the advertisers intended: "You'd have to eat 12 bowls of Cereal B to get all the nutrition in just one bowl of Fibergritz." Quasi-surveys attempt to convince human advertising targets that a majority of experts or fellow consumers prefer the product in question. But the apparently casual statements may mask statistical abuses: "Four out of five dentists prefer chewing sugarless gum."

Within each area—price, product, and popularity—advertisers are almost certain to omit some crucial piece of information without which their claims are more or less meaningless. The essence of advertising mathematics is the logical vacuum, a void in which a single number or fact floats without any frame of reference. Logic is useless in such cases; you just know something is missing. In fact, not even the logic of the famed Mr. Spock, that paragon of extraterrestrial intelligence from the "Star Trek" television series, can help. Especially if, like Spock, one makes a cosmic logical blunder.

In one episode of "Star Trek," the starship *Enterprise* was hit by an ion storm and the power went out. Captain Kirk wondered what Mr. Scott, the engineer, was up to. Spock replied, "If Mr. Scott is still with us, the power should be on momentarily." Moments later, the ship's power came on and Spock arched his Vulcan brow: "Ah, Mr. Scott is still with us."

Here is a classic logical error. From Spock's own premise and from the sudden reappearance of power, no one can logically conclude anything. Had Spock said, "If the power comes on then Mr. Scott is still with us," his deduction would have been appropriate. As things stand, with the sentence reversed, Spock could not have deduced anything unless Scott had appeared while the power was still off. In that case, he could have arched his other brow and said, "It appears that the power will be on momentarily."

The analysis may sound picky, but who can resist being picky when the pickiest logician in the galaxy fouls up. And if Spock (who represents Hollywood's best effort at being logical) can't get it together, what should we expect from television commercials?

Many television ads seem, at first glance, to carry the day with irrefutable logic, especially when announced by an authoritative voice. But if the mathematical ideas underlying their claims are probed a bit, they dissolve into thin air.

Many of television's most famous commercials embody logical vacuums: Ivory Soap is 99 and 44/100 percent pure. Pure what—soap? And what is the impurity? Could it be dirt? Essence of poison ivy? The ad doesn't say. Would you drink water that is 99 and 44/100 percent pure? Not if you thought the impurity was strychnine.

Rolaids consumes 47 times its own weight in excess stomach acid. On the surface this sounds impressive because it sets up a numerical relationship between the weight of a Rolaids tablet and the weight of the excess stomach acid it consumes. In reality, it's no more meaningful than saying, "Aunt Mary weighs 47 times as much as her cat." Aunt Mary may or may not be overweight. It all depends on the weight of her cat. There simply isn't enough information given in the ad to judge whether "47 times" is impressive or not. You should not be impressed by the Rolaids ad until you know at least two things: How much does a tablet weigh? What weight of "excess" stomach acid will produce the symptoms that require a Rolaid (or two or three)?

Logical vacuums not only omit numerical information, they also omit logical details. Think for a moment about the logical possibilities implied by the following statement in a commercial for Kellogg's All Bran: "Some studies suggest a high-fiber, low-fat diet may reduce the risk of some kinds of cancer." This sounds pretty scientific, hedged as it is with "may" and "reduce" and "some." Kellogg's wants you to think, perhaps, that a diet high in fiber reduces some people's cancer risk, although not necessarily everyone's, a kind of get-healthy lottery. But from a strictly logical point of view, the "may reduce" might apply to everybody, and only in the following sense: "It may reduce the risk of cancer or it may not." You can make such a claim about literally any substance that is not known to actually cause cancer. Cold water would certainly reduce the risk of cancer, especially if you drank it instead of a plutonium cocktail. Also, the "reduction" might be so small as to be unnoticeable. And what forms of cancer does the word "some" embrace? Cancer of the liver or a benign neoplasm? Suddenly a statement that sounded impressive seems absurd and banal.

A Game for the Numerate

In their quest for the perfect vacuum, television commercials that strive to impress so often leave out key items of information that numerate people can enjoy "TV Howlers" almost anytime they watch. Different versions of the game already exist, from merely laughing at the abuse with friends present, to the slightly more elaborate version that abuse detectives play.

The Pepsi Challenge in the previous chapter paved the way for this game. There is only one rule, "Truth in advertising." The statements the commercial makes about the product are assumed to be true to begin with, even if meaningless—like Ivory Soap being 99 and 44/100 percent pure. The player then adds missing information that (a) is consistent with the given information and (b) allows him or her to come to a definite conclusion about the product. If the information you add does not complement the product, that's hardly your fault. You're simply trying to fill in some of the gaping holes left by advertisers. Here's a simple example.

You are comfortably watching a commercial when your intelligence is assaulted by the statement, "Four out of five dentists surveyed recommend sugarless gum." Instantly you reply, "Only five dentists were surveyed!" The statement is perfectly consistent with the advertiser's claim. And a possible abuse is then apparent: The sample size in the quasi-survey may have been too small to draw a statistically valid conclusion from.

A more complicated example involves the same logical vacuum as that in the Rolaids commercial. But the stakes are higher because the numerical and logical relationships are definitely more complicated. The announcer says, "You'd have to eat four bowls of Raisin Bran® to get the vitamin nutrition in one bowl of Total®."

Where's the TV Howler here? Take it step by step and you'll have the answer before regular programming resumes. First, spot the operative words and phrases in the commercial. These include "four bowls" and "the vitamin nutrition." The latter phrase seems to imply "all the vitamins" and the truth of this claim might depend on a single vitamin. Suppose, for example, that Raisin Bran is actually far superior to Total. It contains four times as much vitamin A as Total, four times as much vitamin

B, and so on. But perhaps there's one vitamin that Raisin Bran actually contains less of than Total.

Since the commercial doesn't mention any missing vitamin, you may suppose it to be some minor vitamin like beta carotene (plentiful in carrots). Suppose, then, that Total contains four times as much beta carotene. To get *all* the "vitamin nutrients" in one bowl of Total, in fact to get four times as much, you would merely have to eat *one* bowl of Raisin Bran with a few sliced carrots (for beta carotene) added.

If the superiority implied by the four-bowl comparison depended on the absence of a single ingredient in the competing product, the TV Howler turns the tables on the advertiser by pointing out that, in all other respects, the product being advertised may be inferior to its competitor. The real question is this: Was the advertiser trying to conceal some weakness of the cereal behind the four-bowl bravado or was it simply unaware of how little it really said about the product?

Ivory Liquid Goes to Camp

Television is a medium of more than just words, of course. It sprays us with a barrage of consecutive pictures, each worth the proverbial thousand words. Its visual power gives it ample scope to commit the electronic equivalent of chart abuse. In Chapter 1, geometric perspective made the malignant melanoma chart look much worse. The people at Ivory used the same trick to make their dishwashing detergent look much better than a competing brand. It represented two numbers by the lengths of two rows of plates.

The setting was a summer camp. A hundred kids had just eaten from an equal number of plates at an incredibly long table. Before being allowed to run off to torment lower life forms, the kids were made to wash all the dishes. What better time to compare Ivory Liquid with its closest competitor?

The plates washed by one capful of Ivory Liquid were arranged in a row along the south side of the long table, and the plates washed by a capful of the competing brand were laid along the north side. After panning down the table, the camera showed a scene not unlike the one shown in Figure 5.

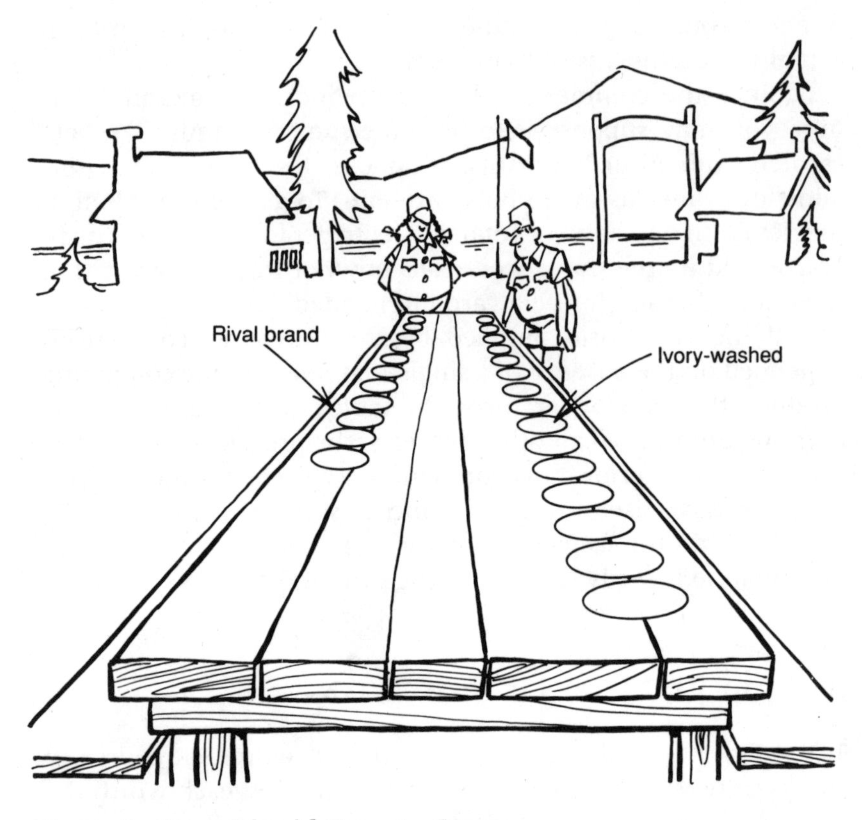

Figure 5 Ivory Liquid Goes to Camp

Off in the middle distance, but looking inexpressibly tiny, was the last plate in the competing row. But there, running right up to the camera on one side of the screen, were the plates washed by Ivory Liquid. The diminishing power of perspective made it seem to the eye that the Ivory had not washed a mere 30 percent more plates, as claimed by the announcer, but about ten times as many.

Had the scene been shot from the other end of the table, you would see almost no difference between the two detergents; each row of plates would stretch into the middle distance and each row would appear to stop at the same spot. There really isn't much difference between detergents, a conclusion that kids at science fairs have been arriving at for years.

One-Page Wonders

The print medium may have no sound and no moving image, but it was used for advertising long before television came on the scene. In fact, print ads contain images and words, as well. They just don't move.

Although typical printed advertisements have only a tiny fraction of the potential information capacity of television commercials, they seem able to deliver essentially the same messages and just as many abuses. The nice thing about print is that it stands still. You can ponder its logical vacuum long after a television commercial would have ended. You can admire the skill and ingenuity that must have gone into the research for the ad. You can even remove it from its page and frame it.

Take the case of the Cutty Sark ads that ran in *The New York Times* during the late 1980s. In blind taste tests, a surprising *half* of non-Cutty Sark drinkers preferred the taste of Cutty Sark. The pitch is nearly identical to the one used by Pepsi (see Chapter 2), which claimed that in recent blind taste tests *more* Coke drinkers surveyed preferred the taste of Pepsi.

Not ready to trash unfavorable tests in order to claim more than a 50 percent preference, the Cutty Sark advertisers made a more modest claim: Fully half the drinkers tested preferred the flavor of Cutty Sark. Perhaps the well-known whiskey is truly superior, but perhaps none of the tasters could tell the difference between Cutty Sark and the test brand.

Will Life Jackets Kill You?

A more elaborate ad is from Sweden, but it's a classic of a logical vacuum in print. The missing information is actually so important that, as in the All Bran commercial, its presence might actually undermine the ad's message.

Most people believe that boating without a life jacket is dangerous. But if they didn't believe it, this public service ad shouldn't necessarily convince them:

SIX TIMES AS SAFE!

The ad appeared in Swedish newspapers. A rough translation of the main message goes something like this:

> LAST YEAR 35 PEOPLE DROWNED IN BOATING ACCIDENTS. ONLY FIVE WERE WEARING LIFE JACKETS. THE REST WERE NOT. ALWAYS WEAR LIFE JACKETS WHEN BOATING.

The ad encouraged people to draw a striking conclusion from these data alone: Some 30 of the drownings (about 85 percent) were presumably caused by missing life jackets. The other 5, some 15 percent of the victims, wore life jackets but drowned anyway. Since six times as many of the victims did not wear life jackets as did, the author of the ad claimed that wearing a life jacket is "six times as safe." In fact, the six-times figure is something of a Baltic herring, so to speak.

The real question is this: What percentage of Swedes wear life jackets when they go boating? The answer to this question can make a very big difference in how we interpret the numbers in the ad. For example, suppose that as a rule fewer than 15 percent of Swedes wear life jackets. In that case, a higher proportion of drowning victims wore life jackets than did the general population of boaters. This could only mean that, from a statistical point of view, it's dangerous to wear a life jacket.

Naturally, if more than 15 percent of Swedes habitually wear life jackets when boating, the lower proportion of drowning victims wearing them would be a direct clue to the efficacy of life jackets in saving lives. If this is the case, the writer of the ad would have done better to include the information, perhaps like this:

> LAST YEAR 35 PEOPLE DROWNED IN BOATING ACCIDENTS. ONLY 5 WERE WEARING LIFE JACKETS. *MOST BOATERS WEAR LIFE JACKETS.* ALWAYS WEAR YOURS!

Presumably, it's true that most Swedes wear life jackets. If so, *more* than 15 percent do and this, in turn, implies that life jackets work because a smaller percentage of the population that drowned wore life jackets.

For Sale: High-Price/Low-Performance Computers

Here, as in earlier chapters, public service messages go awry as advertising types struggle with elementary math and end up losing. You can be more forgiving when the ostensible aim of the ad is to save your life. But if the purpose of an ad is to sell you a computer, it had better have its mathematical act together.

What on earth was on the collective mind of a certain computer company when it advertised how awful its computers were? Imagine the company's embarrassment when it realized that a slick ad placed in several trade publications stated exactly the wrong statistic to impress potential customers. Since everyone in the computer industry (with the possible exception of senior executives) is numerate, you may have to assume that the blunder originated in the advertising department.

The ad showed a series of vertical bars comparing the price-to-performance ratio of their computers with those of its competitors. Guess who had the highest bar. Guess who was proud of it!

A moment's reflection explains why, instead, the company should have been embarrassed. To calculate the price/performance ratio, simply divide the price of a computer by a number that reflects its performance, say the number of operations it will perform in a second.

$$\frac{\text{price}}{\text{performance}}$$

There are two ways to produce a high ratio: Increase the numerator (price) or decrease the denominator (performance)—or do both things at once. To have the highest such ratio means either that your computers are higher priced than those of your competitors or that their performance is worse, or perhaps both. In a numberscape where bigger is always better, the innumerate sooner or later pay a heavy price for their own poor performance.

The 226.8 Gram Canary

Another abuse occurs all too commonly in advertisements: the pointless precision of dangling decimal digits. Numbers with a decimal digit or two thrown in often impress people, but frequently the extra digits are unnecessary. Sometimes they're downright silly.

There was nothing wrong with the glossy camera ad on the back cover of the photography magazine. It showed a remarkably clear shot of a rare bird. Accompanying the image was a brief, scientific summary of the creature, all very impressive, including the average adult weight: 226.8 grams.

Did the decimal digit matter? Where did it come from, anyway? A moment's work by the abuse detective who caught this one revealed a clue. Might the weight (in grams) have originated in another system of measurement? Hmm, try pounds: A quick check in a table of weights and measures revealed that one pound equals 453.6 grams. Amazingly, this figure worked out to exactly double the scientific-sounding weight for the rare bird. Its average adult weight is a half-pound.

Surely that figure is approximate. Not only the decimal digit but the third digit of the weight were completely superfluous and misleading. Have scientists weighed every single living specimen, then taken a complete average to the nearest tenth of a gram? Not likely.

There is a common confusion about numbers with a multitude of digits. How many is enough? If a number carries too few digits to be really informative or useful, it becomes what I call a *num*. The missing "ber" represents the absent digits. Nums can be misused by companies, for example, to charge you too much or deliver too little.

The opposite abuse, that of carrying too many digits for the sake of impressing people, like the 226.8 gram canary, is found frequently. I call the excess digits, on whichever side of the decimal point they occur, *dramadigits* because they lend a certain drama to any number they accompany. The more there are, the more precise or "scientific" the answer appears to be.

Mr. Spock, to harken back to television's paragon of logical thought, throws in dramadigits whenever he can. This practice lowers my respect for him still further. Once, for example, Captain Kirk and Spock were hiding out in an encampment of Klingons, dreaded enemies of the Federation. Kirk wondered out loud if they would ever escape, and Spock replied with something like, "The probability of our eventual escape is only 0.000162, Jim." Not much better than the chance of winning a lottery, which we shall explore further in the next chapter.

4

Intelligent Dice

Gambling, or the attempt to get money without earning it,
is a vice which is economically . . . ruinous. In extreme cases
it is a madness which persons of the highest intelligence
are unable to resist.
—George Bernard Shaw, *The Vice of Gambling and the Virtue of Insurance*

Well-recognized vices such as smoking, drinking, and (especially) drugs have elements of addiction and self-destruction in common. Psychologists also recognize gambling as an addiction and, in extreme cases, consider it just as self-destructive as the other vices. Shaw knew what he was talking about.

Highly intelligent people are not necessarily numerate, any more than they are necessarily literate, less so, in fact. And in games of chance, even numerate people may fall prey to curious beliefs. Roulette wheels, cards, and slot machines foretell the future. Dice become intelligent. How else can you explain people's belief in the so-called *law of averages?*

In gambling, lotteries, and even game shows, the desire to win focuses undue attention on statistical anomalies such as runs of good and bad luck. When gamblers experience a thrilling sequence of wins, they say they are having a hot streak, but fear that their luck will change. On the other hand, when they have been losing heavily, they are sustained by the belief that their luck is about to "turn." They believe that the longer they wait, the more likely it becomes that the next roll of the dice will favor their bet. Somehow, the dice must know what has been going on. Somehow, the dice roll in a way that accommodates the expectation. Or do they?

Even in the seemingly innocuous lotteries, the same behaviors may emerge. Some lotto players have no other game plan but to wait for the big win that will jump-start their lives. The dream of winning may sustain them in some sense, but it also erodes their ability to formulate and pursue realistic goals. They pay double for the dream. They lose money and they lose opportunities.

Dr. Lotto Dies Under the Knife

Dr. Lotto makes big promises on his prominent poster displayed on an uptown bus. By calling a special telephone number, paying $1.50 for the first minute and 75 cents for each subsequent minute, you will learn what numbers have already won lotteries. Some lottery players call these numbers "hot." For example, if 14 cropped up in four out of the last five state lotteries, 14 would be considered hot. Other numbers, having won heavily in the past, may not have appeared recently. They become "due."

Avid lottery players have developed a host of strategies for picking numbers. Most strategies have no more chance of success than the standard practice of using your birth date or other "personal" numbers.

Some of Dr. Lotto's colleagues live in computers—not people, but programs you can buy. They track hundreds of winning numbers and dates and produce "expert trend charts" not only for hot numbers, but for ones that are due.

It would be easy enough to point out the utter waste of time it is to expect past numbers to predict future ones, or to explain that charts are purely illusory. But we can operate directly on Dr. Lotto by cutting straight to his heart, the lottery machine itself. My favorite is a French model that generates numbers for 6-49-style lotteries. With it, I can show that the combination 1, 2, 3, 4, 5, 6 (which many people think would be a sure loser) is just as likely to win as 7, 12, 19, 28, 37, 41, a combination of currently hot numbers.

Everyone has seen a lottery machine on television at one time or another. It looks like a surreal corn popper. A current of air blasts up through a rotating plastic sphere, keeping a bunch of balls whirling about in the most chaotic and unpredictable fashion. The balls are numbered 1 to 49. At a certain moment, one of the balls falls into a collector trough that conveys it to a rack outside the machine. This is the first number. But what number will it be? Will it be a hot number? If so, how does the machine know about hot numbers? Does it somehow tap into a kind of numerical *ambience?* Mon Dieu! All numbers are equally likely.

Which ball ends up on the rack is determined by many factors: the initial arrangement of the balls before the machine was turned on, the current of air that blows up through the sphere, and, especially, the thousands of physical interactions between the balls as they bounce around inside. These interactions are beyond the power of even the largest computers, programmed with the most sophisticated simulation software, to predict. The new field of chaos theory tells us that the interactions are *inherently* unpredictable. No system, no theory, no computer, no amount of thought can predict which ball will land in the trough. The number that results from this extraordinarily complex process is as random as anything we know on earth.

Even if a true believer in hot numbers were to say, "Aha! You mean it might not be random at all!", I would reply that there isn't the ghost of a chance that the number chosen by this process has anything whatever to do with the numbers chosen by similar (or different) machines on other occasions. The probability of any one of the 49 balls emerging to the outside world is always exactly 1/49.

What confuses many people is that when a *particular* ball emerges, as it must, the probability becomes particularized, so to speak. Norming takes over as particular qualities of that number wipe out any realization that it arrived by a random process. For example, if the number is 12 and the previous number was 11, some people will suspect that the process isn't random at all. Other people believe that the ball was destined to emerge and that destiny itself has a pattern that Dr. Lotto might discover. Whether or not the balls are predestined, they are forever beyond any system to discover. As far as ordinary human knowledge is concerned, all balls are equally "destined"!

Here comes the first ball now. It is number 1. This is not amazing. It could just as easily have been ball number 28. And now, here comes the second ball, number 6. It could just as easily have been 41. The third, fourth, fifth, and sixth balls arrive in quick succession: 3, 2, 5, 4. They could just as easily have been 19, 12, 37, and 7. Sort the two combinations out into standard, increasing order and you get 1, 2, 3, 4, 5, 6 and 7, 12, 19, 28, 37, 41. In this particular case, the first combination was chosen. Next time, it could just as easily be the other combination's turn.

There is a particular reason that most people believe that 1, 2, 3, 4, 5, 6 is not as likely to occur. It is memorable. The sequence 7, 12, 19, 28, 37, 41 is not. Close this book immediately and try to repeat both sequences. You will remember only the first one, even though I have repeated the second one four times. Not surprisingly, most winning lottery numbers have this nonmemorable quality. After all, only a tiny fraction of all six-number combinations has anything memorable or remarkable about it. By a simple application of what logicians call inductive reasoning, the ordinary person concludes, in effect, that memorable numbers are not good bets for lottery plays because memorable numbers never seem to win. Inductive reasoning has its place, but not here.

We humans come equipped with a strong inductive bias in our thought patterns. This bias may have kept us alive over the millenia. For example, most of your ancestor's friends who enjoyed petting large animals never returned from the shade of the acacia tree where the lions were resting. Consequently, your ancestors didn't feel strongly tempted to pet lions themselves.

Not surprisingly, instinctive thought biases can be difficult to set aside in favor of rational truths. A high-school physics teacher once asked his students if they believed in the law of conservation of energy. They did. The teacher took them to a gymnasium where he had suspended a 100-pound iron ball by a cable from a ceiling beam. He invited one of the students to stand on a box some distance from the ominous iron ball. Then he slowly swung the ball laboriously up to within an inch of the student's nose. He released it and quickly ducked out of the way. The ball swung ponderously away from the student and over the gym floor, gathering speed. "Do you believe in the law of conservation of energy?", the teacher asked, as if repeating a catechism. The student said he did. The ball returned with grand velocity straight at the student's head. With a yowl of terror, he dived from the box just as the ball came to hover, momentarily, an inch or two from where his nose had been, an inch short of where the ball had started its swing.

The point of the story is that it takes a certain resolve to master instinctive knowledge. We "know" instinctively that 1, 2, 3, 4, 5, 6 is not as likely as whatever that other combination was. By the same kind of inductive instinct, we also "know" that numbers that have been part of winning combinations in the past are more likely to occur again. We may even "know" that our chance of winning a lottery is "slim." But, as I pointed out in Chapter 2, it would be much closer to the truth to say that we have no chance at all of winning.

Winning Ways?

The real lottery winners are the lottery corporations themselves. Many of these corporations enjoy revenues of over a billion dollars a year. The money that *doesn't* go to prizes and overhead (about a third of the revenues) is spent on a variety of projects, some of them worthwhile. But even with this moral sop, lotteries may actually do more harm than good in the long run. They are politically popular because both politicians and a large segment of the electorate love them. They work like a voluntary tax. Contributions to this tax are levied by people on themselves! And many can ill afford to pay it.

You can easily figure out your expected winnings in a six-number lottery. Simply multiply the amount you might win by the probability of winning. Then multiply the amount you might lose (your bet) by the probability of losing. Subtract the second number from the first to get your "expected" winnings (in the statistical sense). I will explain why this is a reasonable way to figure your return in the last chapter.

For example, suppose you buy a $1.00 ticket in a six-number lottery. The jackpot is currently $1 million. If you win the lottery, you win $999,999, or $1 million minus the cost of the ticket. If you lose the lottery, you lose $1.00. When you multiply each of these numbers by their respective probabilities and subtract the second figure from the first, you get

$$\$999{,}999 \times 0.000000013 - \$1 \times 0.999999987$$

or about *minus* 98.7 cents. Frankly, that's what the gambling crowd calls a *sucker bet.*

Even if most lottery players do not appreciate the continuing erosion of their money at the rate of $0.987 per play, let them at least consider what would happen if the long-awaited event should actually come to pass. The disillusioning news will be spelled out shortly.

Are there strategies that will help you win lotteries? There is only one strategy that will enhance a lottery player's winnings, but I hesitate to give anyone even the slightest encouragement. I can't help feeling that it will backfire on me somehow. The simplest scheme might be called *playing the players.* If all number combinations are equally likely, you may as well avoid numbers that other players are likely to use. That way you won't end up sharing the prize in a multiple-winner outcome.

The avoidance strategy has an immediate application: *Don't* play Dr. Lotto's numbers. Too many other people will be using them. Unfortunately, this means you will have to buy Dr. Lotto's services in order *not* to use his numbers!

Winning Big

Set aside schemes for winning and suppose, just suppose, that you are one of those rare, extremely lucky individuals who actually wins. What will you do with your million-dollar prize? Some lotteries, like Canada's 6/49, pay all the money out at once. And some prize winners, like the unfortunate Montreal woman, whose story was told on a recent Canadian Broadcasting Corporation public affairs program, are simply not ready for all that money.

According to the program, the woman recently won $1,199,168.68 (Canadian dollars: $1 CD = 78 cents $US) in Lotto 6/49. The money couldn't have come at a better time. Her jewelry business was in trouble, and she was broke, in debt, and behind in her rent. She paid off her debts to the tune of $323,000; spent $160,000 on a new home with a swimming pool, clothes, and jewelry; loaned or gave away hundreds of thousands of dollars to relatives; and began to throw lavish parties regularly. It only took her a year to blow her entire winnings.

Unfortunately, things didn't stop there. Her new life-style had to be sustained. She borrowed until she was as far in debt as she had been before her big win. Then she launched a number of lawsuits against relatives in a vain attempt to get some of her money back. Today she is struggling to renew a shattered life with regular visits to a debt counselor.

If you should happen to win a million dollars in a more typical lottery, the first thing you must do is realize that you won't get it. Instead, you will get something like 20 annual payments of $50,000 each. And because you are now in the top income bracket, you will pay a hefty rate of tax on this money. And that is not all! Inflation will eat into your prize. At average rates of inflation (say, 4.5 percent a year*), the purchasing power of that $50,000 will have eroded to something like half its present value. Time may mean money but, in a lottery context, time means lost money. Math mischief is afoot!

Revenge of the Dice

Two gamblers board the plane for a junket to Las Vegas, laughing, joking, and already a little tipsy after visiting the airport bar. Fellow passengers can hear them bragging about their winnings of a few weeks ago and how they expect to win much more this time. Several of the passengers, who are keen to try the famous gambling facilities, lean over in their seats to catch the jargon.

"That was three big wins at blackjack and one on the slots. I tell ya, Dave, if we could have stayed just one more day, I would have beat the roulette at Caesar's. I almost picked up a grand on one bet!"

*This is a fairly good approximation of the long-term average rate, in spite of the fact that inflation has recently been somewhat lower.

In truth, the two of them have lost far more than they have won, but good sense does not rule Vegas; false hope and blind greed do.

Math abuse and innumeracy meet with a vengeance in the casino. Consider the slot machines. A roomful of these makes a more or less continuous ringing sound, as John Paulos, author of *Innumeracy*, has pointed out. Suppose, for the sake of argument, there are a hundred of them. If each machine pays off just 1 percent of the time (a handful of quarters is typical), then, on average, you will hear one paying off every 10 to 20 seconds. This is the sound of winning. Losing, of course, makes no sound at all. But imagine what the room would sound like if the machines were rewired to say "Woe is me!" every time they failed to pay off.

Slot machines account for about 40 percent of casino industry "earnings." According to John Scarne, a longtime observer of the gambling scene, the simplest machines pay back about 90 percent of their take in prizes, leaving 10 percent—the "house advantage"—for the owners. In other words, in the long run, you might expect to lose 10 percent of your capital, more or less. If so, this levy is relatively mild compared to lotteries where you expect (by the expectation formula) to lose 99 percent of your money.

Like the lottery corporations, the "house" or gambling casino is the big winner in virtually all its operations. The following passage from *The Eudaemonic Pie* by Thomas A. Bass illustrates the point very nicely:

> You see the casinos wheeling carts around to pick up the cash boxes. Hour after hour they're raking it in. You're sitting next to losers who don't leave the table until they're broke. You see their wives pulling on their sleeves, saying, "Honey, don't do it. That's our bus fare." And then it's gone and you have no idea how they're going to get back to wherever it was they came from.
>
> The gambling is all run by the Mafia. And it's their best business. Who knows what makes people do it? It's an animal instinct, an atavistic trait, a disease. . . . It's naked reality. I mean the casinos are so bad and greedy—screaming about every nickel they lose, while cheating

> and robbing people blind—that to beat them at their own game is a white knight operation.

Slot machines are not the only games to give suckers a less than even break. The blackjack tables enjoy a mere 5.9 percent advantage with unsophisticated players. A certain type of gambler takes the challenge more to head than to heart. He or she will attempt to narrow the advantage by counting cards or playing an exotic strategy (which may be bogus or genuine) to cut the house advantage to as little as 3 percent or even less.

The craps table yields the house an advantage of anywhere from 1.4 percent to over 11 percent, depending on the type of bet. The roulette table takes 5 5/9 percent for the house, on average. From all of its games, the take for the house is slow, steady, and statistically inexorable. If the gamblers do not notice the continuing drain on their pockets, it is because they are lost in the innumerate dream of instant money, hurried along by an addictive rush of adrenaline. The most common misconception has a common name: the law of averages. I call it "the law of creeping chance."

It works like this: A gambler watches the roulette wheel, waiting to bet on whether the number that comes up will be even or odd. (Figure 6 shows a roulette wheel and table layout.) She places a chip on the ODD rectangle and waits. The croupier gives the wheel a spin and sends the ball racing around the track until it tumbles into one of the numbered cups on the wheel. If the number the ball lands on is odd, the gambler collects on her bet. If the number is even, however, the croupier rakes in her chip.

When the ball lands on an even number, the gambler grimaces in exasperation and plays ODD again. This has been going on for seven plays in a row. "This time," she says to herself, "the ball will almost certainly land on an odd number. It's the law of averages."

In truth, the ball is no more likely to land on an odd number now than it ever was. But the law of creeping chance (otherwise known as the law of averages) can be stated as follows:

> The longer you wait for a certain random event, the more likely it becomes.

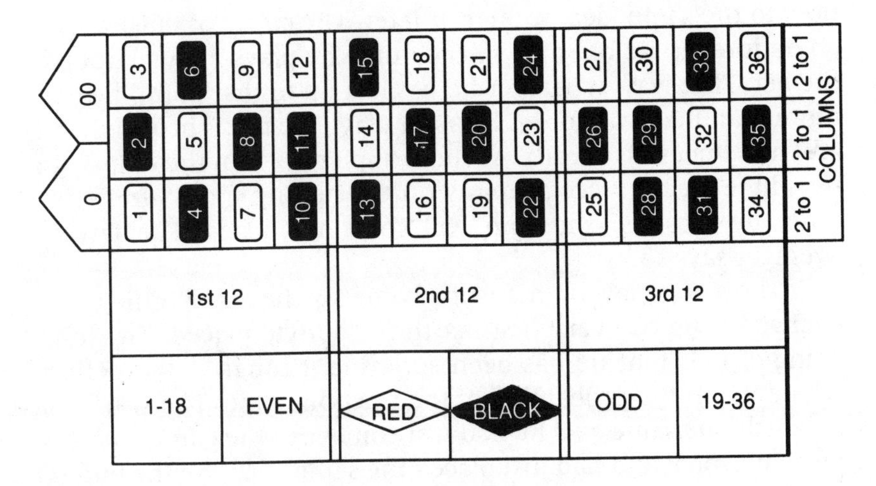

Figure 6 Roulette Wheel and Table Layout

It is quite wrong. The reason so many people believe in the law of creeping chance may have to do with a confusion between the probability of a run and the probability of a single event in that run. I can illustrate the confusion with a simplified casino game you can play with a coin. I call it "flipout."

The dealer flips a coin and a neophyte gambler bets on the outcome. By the rules of the game, he sets the amount of the bet. If the coin comes up heads, the house pays him that amount. If the coin comes up tails, he pays the house the same amount. He enters the game with a certain number of $1.00 chips. Should he lose them all, he will be "flipped out" and must retire.

Suppose he begins a game of flipout and loses the first toss. Now down a dollar, he tries a second time and wins his dollar back. Encouraged, he tries a third time and wins again. Suddenly, he grows reckless and bets two dollars. My goodness! He wins again. He grows flushed with a strange pleasure he has never felt before. It is called "gambling fever."

Before long, he starts to lose. A string of five tails in a row reduces his capital to the level he started with. That's when he hits upon a very simple scheme. Every compulsive gambler falls prey to the same idea, sooner or later. The string of tails cannot go on forever, he reasons, so each time he plays, he will double the bet. That way, whatever he has lost up to that point, he will more than make up for on his next play. The next throw is a tail and he catches his breath. As the croupier rakes in the chips, the neophyte places twice as many on the betting spot. Surely this time (he reasons) the coin will come up heads. The law of averages says so.

The coin may or may not come up heads. Nothing has changed from the very first bet the neophyte placed. The long string of tails that he has been suffering through has no effect whatever on the probability of the next outcome. The situation is exactly the same as if he had just come into the game with his present capital and had just placed the same bet. Like the lottery machine and the roulette wheel, the coin knows nothing at all about past history.

What the gambler thinks he is betting on is that the long run of tails must, sooner or later, end. That is a safe bet, as I will show in the last section of this book, but that is *not* what the gambler is betting on. He is simply betting that the next toss will be heads. If it isn't, he will lose his pile of chips. If it is, he will no doubt experience an even greater rush, completely lose his head, and begin to play all over again. Until, like any compulsive gambler, he flips out—in both senses.

A mathematical paradigm, taught to all math majors in statistics classes, is called *the gambler's ruin*. It involves a demonstration that, sooner or later, as long as the gambler keeps playing, he will lose all his capital. If only he could be taken aside at some point and shown the coin-toss chart (Figure 4, Chapter 2). Short runs of tails are fairly frequent; very long runs are infrequent. But there is no real upper limit to how long a run may be. If the gambler could play forever and have infinite capital, he would, with "probability 1," experience runs of all possible lengths. This can only mean that if he starts with any finite amount of capital, he will sooner or later place all his remaining chips on the table, lose that last bet, and go home broke. In this context, "probability 1" means that he will have to be extraordinarily lucky *not* to suffer this fate.

The problem with doubling the bet is that it involves exponential growth in the amount the gambler must place on the table each time he loses a bet. As you may remember from Chapter 1, exponential growth is formidable. For example, ten doublings of a $1.00 bet require $1,024. And runs of ten consecutive tails, though unlikely in the short run, are almost inevitable in the long run.

Those who counsel compulsive gamblers (in the hope of modifying their behavior) know that the bet-doubling syndrome is very common. Since most casino games involve some statistical advantage for the house, gamblers tend to lose their money faster than they would in the pleasant little coin-tossing game I have just described. Moreover, real casinos generally apply a house limit to all bets. The typical maximum bet of $500 on the roulette table, for example, only hastens the gambler's ruin because if he or she loses the most recently doubled bet and the amount is just under $500, it would not be possible to double the bet again.

The abuse and innumeracy in a typical casino go well beyond the law of creeping chance. Some players at the roulette table will split their bets, for example, betting $10 on RED (the wheel numbers have different colors) and $10 on ODD. That way, they reason, they will have twice the chance of winning. That is true, but they may win only half as much. If you want a little practice working out the player's expectation, consult the image of a roulette wheel (Figure 6) and work out the probability of getting RED, the probability of getting ODD, the probability of getting both and the probability of getting neither.

Now, for each kind of outcome, multiply the probability by the amount to be won (positive) or lost (negative). When you add up the resulting numbers, don't be surprised if the expected amount won is exactly the same as a $20 bet on just one of the events.

You walk up to the blackjack table and nudge someone. You want in.

"Wait until the shoe is over!" he growls.

"Why?" you ask.

"Because you'll change the order of the cards!"

The "shoe" is the present deck of cards—as shuffled. By entering the game and being dealt some of these cards by the dealer, the new player will change the cards that *everyone* will get. He thus runs the risk of changing everyone's "luck." He could destroy a winning streak, for example.

The same thing happens if a neophyte player makes a bad play, getting "hit" with too few or too many cards. This now affects what cards everyone will get until the end of the shoe. They glare at the neophyte with genuine and completely misplaced anger.

The more you know about gambling, the less you should want to play. In the end, if you still feel like taking a chance, try playing, "Let's Make a Deal". There, at least, you have a 50-50 chance of winnning a car. Maybe.

The World's Smartest Human Screws Up—or Does She?

Up to this point, I have spoken of numeracy as if it were an all-or-nothing quality. Either you have it or you don't. In truth, there are grades of numeracy and even the most numerate have their bad moments. The best-known of recent cases involved a certain well-known game show called "Let's Make a Deal." It all started in a magazine.

In September 1990, an innocent letter to Marilyn vos Savant, author of the "Ask Marilyn" column in Parade magazine, began a debate that still rages. The question involved a variant of the famous televison game show "Let's Make a Deal." In this show, host Monty Hall shows a trembling contestant three curtains. Behind one is a car. Behind the other two curtains are prizes of somewhat less value, like a year's supply of Handi-Wipes.

Hall asks the contestant to choose a curtain and she does. Hall might even ask the contestant whether she would like to switch. If she says "no," Hall might open one of the other curtains, revealing a set of beautiful Tupperware mixing bowls. He may try to get the contestant to change by throwing in additional cash incentives, bribing her in effect. But what if the car is behind the curtain the contestant has already chosen? The tension has kept millions of viewers glued to their sets over the years.

The letter to Marilyn vos Savant described a closely related game. Ms. vos Savant, widely known as the Guiness Book record holder of the world's highest IQ, must have spent at least a minute pondering the fateful question:

"Suppose you're on a game show and you're given the choice of three doors: Behind one door is a car; behind the others, goats. You pick a door, say No. 1, and the host, who knows what's behind the doors, opens another door, say No. 3, which has a goat. He then says to you, 'Do you want to pick door No. 2?' Is it to your advantage to switch your choice?"

Her Answer? "Switch."

The reason? When a contestant first chooses a door, he or she has a one-third chance of being right, that is, a probability of 1/3. The probability that the car is behind one of the other two doors is therefore 2/3. When the host opens one of the other doors to reveal a goat, the 2/3 probability now attaches, so to speak, to the remaining door. The contestant doubles his or her chance of winning by switching doors.

Ms. vos Savant's analysis of the problem drew a great many letters from academics, physicists, and even a few mathematicians, who with various mixtures of scorn, condescension, and pity, pointed out that after the door was opened, the probability was now evenly split between the remaining two. There was no special advantage to switching. Several of the writers even disparaged Ms. vos Savant's role in spreading innumeracy! Ms. vos Savant, however, was perfectly correct.

The controversy about the three-curtains problem has spread so far and wide that readers of this book may be more or less evenly divided about the answer. Those who agree with Ms. vos Savant will nod their heads sagely while those who do not agree will become quite annoyed with the drift of my exposition.

When mathematicians find themselves faced with such a dilemma, they always go back to the definition. What is the probability of winning if you always switch? In this case, the definition becomes the following fraction:

$$\frac{\text{number of cases where switching wins}}{\text{total number of cases}}$$

Suppose you are a contestant on this game show. You will be given the opportunity of witnessing all possible outcomes and recording them for the sake of this experiment. To simplify matters, you will always make Curtain 1 your first choice. You can attempt the more complicated analysis where any curtain will be your first choice, but it will come to the same thing.

You chose Curtain 1. What are the possibilities?

Curtain	*1*	*2*	*3*
	car	goat	goat
	goat	car	goat
	goat	goat	car

In the first case, your curtain conceals a car and you will lose by switching when the host opens another curtain. In the other two cases, however, you will win by switching. The numerator of the fraction is, therefore, two and the denominator, the total number of cases, is three. The probability of winning by switching is, therefore, two-thirds. The probability of winning the car if you don't switch should now be apparent: You will win it in only one case out of three for a probability of one-third.

Most of vos Savant's indignant letter writers understood the premise of the game as well as she did, that the cars and goats game host will *always* offer a switch. Very few of them had the good grace to apologize or acknowledge their own innumeracy when she published her rather clear-cut argument for why a contestant should switch. The argument was simple. Suppose there are 100 doors and only one has a car behind it. The others all have goats. If you pick a door and the game show host shows you 98 other doors with goats behind them, you'd be a fool not to switch to the one remaining door. It very probably has a car behind it.

Now you know what to do if you ever get invited to be a contestant on the Goats and Cars game show. Unfortunately, the same strategy won't work on the famed "Let's Make a Deal" game show. Host Monty Hall does not always offer to open another curtain to give a contestant the opportunity to switch. As Martin Gardner, longtime puzzlemeister with *Scientific American*, points out, "if the host is malevolent, he may open another door only when it's to his advantage to let the player switch, and the probability of being right by switching could be as low as zero."

How appropriate that we are about to enter the world of finance, both high and low, only to discover a frightening similarity with the world we have just left—gambling.

5

The Law of Zero Return

"A slow sort of country!" said the Queen. "Now, *here*, you see, it takes all the running you can do, to keep in the same place."
—Lewis Carroll, *Through the Looking Glass*

If someone gave you a handful of money, your first question might be, "How much is there?" You would count it, adding up the tens, fives, and ones, until you came to a definite amount. Money has this peculiar property. Any bundle of the stuff always amounts to a specific number. Time may be money, but money is numbers.

People who fear numbers invariably get a bit muddled when it comes to handling money or even thinking about it. Making change at a counter is one thing, but handling thousands of dollars is quite another. If nothing more complicated than adding mistakes (or, at the worst short change) attends small transactions, it takes little to imagine what happens when large amounts of money trade hands. To put it abruptly, other people want your money. One way they get it is to mislead you about the nature of the transaction.

Shortchanged!

Shortchange artists still work the street. According to an officer who works in the fraud squad of a major North American city, shortchange artists still work the street, but reports are rare.

"Who's going to call up just because they got beat out of $25? It's not worth their while. Besides, they're supposed to know better."

The following example of a shortchange artist at work comes straight out of a book on magic. As the standard shortchange classic of all time, it serves as a lesson to magicians and simultaneously illustrates what mental sleight of hand can mean. Strangely enough, the same technique was described to me by an old high-school friend whom I reencountered after taking up my first academic position. We met at a party. He unabashedly confessed to having spent several years playing in a rock band, panhandling, and, when the going got very tough, running a shortchange scam down one block and up another. He could take every third store, on average, for ten bucks at a time. If he needed a few hundred, he would do this for a day. He introduced his description by smiling and saying, "You'll like this, it's almost mathematical."

"I go into a store, right. I pick up something that costs less than a dollar, say, some Rolaids. I put the Rolaids on the counter and I give the clerk a $10 bill. The clerk rings up the sale and returns the change, coins first. As soon as I have all the coins in my hand, I pick up the Rolaids and leave the counter. The clerk calls me back. 'Hey,' she says. 'You forgot the rest of your change!'

"The trick is to look dumb. You know how a certain type of person never seems to know where his change is? I act like that. I go back to the counter and pick up the bills and I say, 'Oh, yes. The rest of the change!' Then I look confused. I feel in my pocket where I keep a $1 bill.

" 'Look, I'm sorry about this, but I've got a dollar on me after all. Give me back the ten and I'll give you a five and, uh, five ones.' The clerk gives back the ten and I give her back the $9 she just gave me. I ask her to count it out. 'Five, six, seven, eight, nine,' she says. 'I need another dollar.'

"Now, remember, I've got the $10 bill in my hand now, and the single in my pocket. I'm ready for the sting. 'Say, you've got nine dollars there. Here's another eleven.' I pull the one out of my pocket, add it to the ten and give the two bills to the clerk. 'Just give me a twenty and we'll be even.' I smile at her. If I picked the clerk carefully, she will probably go along with the trade. There might be alarm bells ringing somewhere in her head but because most people don't want to seem stupid, most won't say anything. I walk out of the store nine dollars richer, not counting the coins and the Rolaids."

The shortchange artist uses correct arithmetic and impeccable logic to bilk the clerk out of $10. The numbers, namely 9 and 11, add up to 20 and the logic (if we put our two sums together, we'll have more) checks out. Only one thing is missing, a correct premise. The shortchange artist throws the sum 9 + 11 at the clerk just when she should be remembering that the $9 is not his. He suddenly changes the direction of the transaction without surrendering the change as he ought to.

Most shortchange artists know exactly what they are doing. They also know it is against the law. So much for pocket change. What about the big money? Is anyone out there trying to bilk you out of your savings, and do they know what they're doing? The short, unsurprising answer is, of course, yes. Everyone has heard of con artists operating on a grander scale. People have bought "prime real estate" in Florida, only to discover they own a piece of swamp. Even the savings and loan crisis of recent years could be considered a scam, at least on the part of some of the failed banks. After all, they put out millions of dollars in bad real estate loans, relying on the government to bail out the purchasers when real estate prices plummeted, the S&Ls went bankrupt, and the directors retired to their yachts.

But the biggest scams of all seem to attract little or no attention. Could this be because not even the practitioners realize what they're doing?

Portrait of the Bull

What will you do with that carefully accumulated retirement or education fund you've been patiently building? Should you invest it in the stock market, a mutual fund, a bank, or in a government savings bond? Be careful! The stock market is a haven of math abuse and innumeracy of virtually all types. And what I call the law of zero return lies in wait, even for less adventurous investments like term deposits.

Wall Street uses two animals to symbolize itself, the bull and the bear. In the collective Wall Street mind, the bull stands for a market in which prices are moving up and the bear stands for a downward-moving market. It may only be an unconscious irony that two animals were chosen to symbolize market forces. Animal behavior seems, at times, to dominate the market as a whole. Consider the bull, for example. It stands for bullish enthusiam, also known as greed.

A vivid portrait of the bull market appeared in a January 1987 issue of a major newspaper. This stunning example of chart abuse purported to show the behavior of the Dow Jones Industrial Average (DJIA), a favored indicator of the behavior of the stock market as a whole, from November 1972 to January 1987 when the Dow topped 2000 for the first time.

In order to convey the theme that the market had been "bullish," an illustrator drew the graph using the figure of a bull's head, from the side view, with the smooth upward slope of the bull's horn depicting a steady increase in the Dow average. A number of specific dates and the level of the DJIA on those days were given as points apparently plotted on the curve of the horn, making a smooth curve of growth appear official. The curve of the bull's horn bore scant resemblance, however, to the actual behavior of the market over that period of time, which rose and fell sharply during the course of those years. Figure 7 depicts on the left the curve of growth in the Dow average as it was plotted in the bull graph, and on the right the actual record of increases and decreases we see if recordings are plotted every six months over the same period. The impressively smooth upward curve depicted in the bull graph was achieved by plotting only recordings that fit the curve of the bull's horn, so that many points along the way were ignored and decreases were not accounted for.

It's all very well for some people to say, "But the illustrator is just using a visual metaphor, indicating a healthy market." The problem is that it was a visual metaphor only to those who understood how the Dow actually behaves. To other people, it was a tempting invitation to gamble and catch a ride on that generous upward trend. But as the graph on the right clearly shows, even a bull market offers a rougher ride than the stock novice would have bargained for.

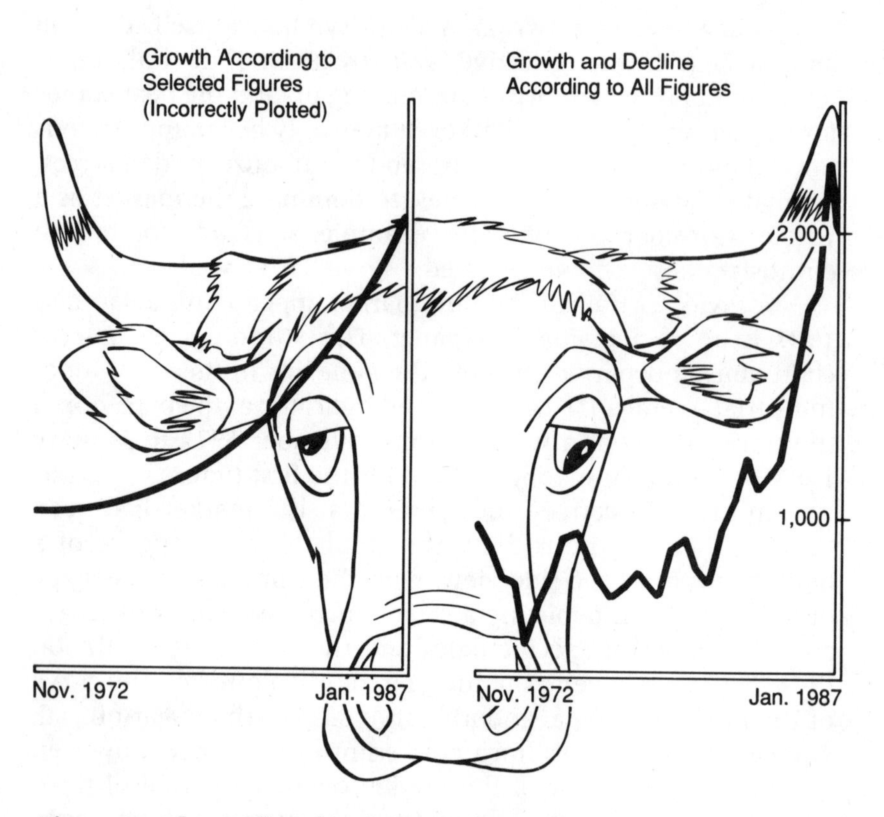

Figure 7 The Bull Behind the Bull Market

Following the Herd

Throughout 1987, up until October, the stock market was hyped by various misconceptions, metaphors, and downright myths, most of them wildly out of line with the actual situation. Speculators were about to find themselves on the horns of a dilemma.

On October 19, 1987, stock prices on the New York Stock Exchange went into a nose dive in a frenzy of selling by panicky speculators and investors. Not since "Black Tuesday" (October 29, 1929) had share prices fallen so frighteningly far or so sickeningly fast. The media all over North America trotted out brokers and analysts to find out what had gone wrong. The gurus of Wall Street and Bay Street fingered government deficits, Japanese competition, rising interest rates—everything but the one possibility that would reflect badly on the purity of the market, their main source of income.

Many "experts" said that the crash heralded a recession, citing past times when this had happened. But as far back as 1960, economist and Nobel laureate Paul A. Samuelson had already summarized the innumeracy of such statements in his now famous remark: "The stock market has called nine of the last five recessions." The imminent recession did not happen. Share prices began to climb once again and, four years later when a recession actually began, stock prices continued to climb with only a few hiccups here and there as the market digested the spare change of the North American economy. In spite of the fact that economists rarely agree on anything, a few actually agreed on the cause of the 1987 crash: A speculative bubble had burst. Poof! To put it another way, the herd had stampeded.

Gazelles, wildebeests, zebra, and giraffes all crowd around the water hole during the dry season. The rank scent of lions nearby makes them nervous but they drink greedily, eyes rolling right and left. When the smell grows stronger, there are nervous whinneys and snorts but still they drink.

Does a lion suddenly rush out of the nearby grass? It might, you never know. But a twig snapping innocently under a roosting bird may be enough. The gazelles bolt, the zebras back up and stampede, the wildebeests moan and bellow, and the giraffes all lope off. The water hole is deserted. What caused it all? A snapping twig was the immediate cause, its precise moment as random as anything on earth.

The stock market works like the water hole. When share prices have been bid up to levels far beyond anything the underlying companies could ever conceivably deliver by way of profits, previously bullish speculators become nervous and look for signs of stampede. They fear not the lion, but the legendary bear. The snapping twig might be new unemployment figures, higher interest rates, even a rumor that a large company has encountered some financial difficulty. Literally anything can cause the stampede, and it always seems to catch true believers by surprise. The phenomenon of market crashes and subsequent attempts by market analysts to hide the truth from others (or themselves) only highlights the innumeracy and abuse that permeate this institution.

The abuses range from the completely pointless drama-digits on the industrial average (3247.83 looks more impressive than 3247) to a belief in chart magic and other predictive schemes. I must warn that the revelations to come will not be immediately obvious, even to numerate people.

Charts and Chicken Entrails

The stock investment industry depends on one major misconception for its continuing health: *It is possible to predict the motion of stock prices.* As long as people with money to invest believe this, they can be encouraged by brokers to part with their money.

The innumeracy of this belief is supported by two basic approaches to predicting stock prices called *technical* and *fundamental analysis.* Technical analysis holds that the future price of a stock can be predicted on the basis of previous prices alone. Fundamental analysis holds that the future price can be predicted on the basis of a company's past and present business statistics.

In his popular book, *A Random Walk Down Wall Street,* Burton G. Malkiel, dean of the Yale School of Organization and Management, points out the uselessness of various schemes for predicting market prices. Although he admits that some forms of fundamental analysis may predict prices with a small degree of statistical consistency, he allows no back door to technical analysis.

Technical analysts, known in the industry as "chartists," believe that everything of importance about a stock can be learned from an examination of its past price behavior. The technical analyst looks for trends, trading channels, and exotica like the configuration shown in Figure 8.

Daily prices, indicated by points, march across the chart, tracing, as they go, the outline of a head and shoulders. If the price completes the right shoulder, the technical analyst inter-

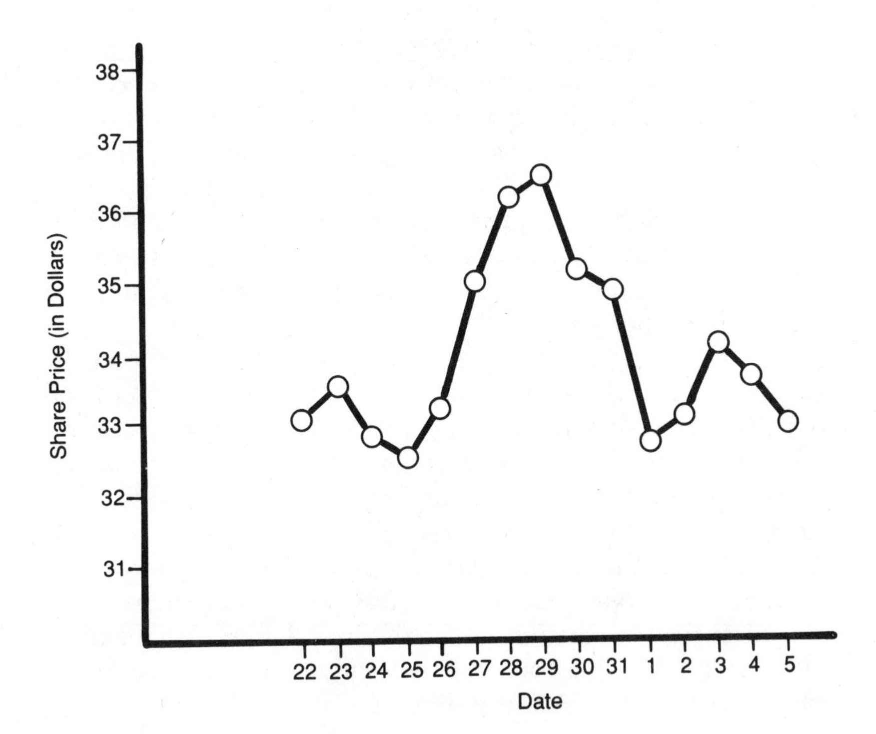

Figure 8 A Typical Head-and-Shoulders Pattern

prets this as a bearish signal and recommends that the stock be sold: The price is about to go down. Why? Who knows? Professors of finance with nothing better to do than to check the claims of Wall Street analysts find that when the method is applied to historical stock price data, it works about half the time. The time-honored divinatory method of reading chicken entrails also works about half the time.

Malkiel and other disinterested investigators with a scientific approach to the equities have formulated the so-called efficient market hypothesis: Prices always move in a way that takes account of every piece of news or information instantly. It might be more accurate to hypothesize that the market does not simply "adjust" to information, it destroys it. But that is a quibble. In any case, no one (except inside traders) knows the news in advance, so no one knows which way the price will react. But even inside traders can be gored by the market's peculiar habit of moving opposite to expectation.

Did I say "inside traders"? Opportunities for this abuse abound because, at the center of the market, pricing decisions are based purely on the pattern of orders to buy and sell stock. If enough people buy a stock, the price will climb. If enough people sell it, the price will drop. The question of what prices will do then is largely a question of how people, with all their hopes, fears, and beliefs, will behave. Those with the ear of many investors and speculators have a unique opportunity to profit from the market.

The following amusing story may be only apocryphal. A broker mails letters to 1,000 of his firm's clients. In 500 of the letters, he says, "Our research department has recently discovered that United Suspenders is about to stretch upward. I strongly recommend that you buy into this snappy oportunity." To the other 500 clients he sends a letter that states just the opposite. When the stock makes a significant move one way or the other, the broker sends out letters to the 500 clients who received correct predictions. In half these letters, he makes a further prediction that the stock will move one way and in the other half he predicts just the opposite.

The broker continues this game for a few more cycles until he has a handful of clients who have received nothing but correct predictions and who, consequently, have become putty in his hands. They have no idea that the broker has filtered out the data they really need to make a correct decision.

Self-Fulfilling Prophets

The real truth is actually worse than this. Brokerage houses, market letters, and even newspapers find themselves in a position to actually influence prices of specific stocks. In most cases, undue influence on the price of a stock, with a subsequent opportunity for profit, may be prosecuted by the Securities and Exchange Commission as a violation of insider trading rules. But you have to catch them at it.

Brokerages, for example, might quietly buy into a specific stock over a long period of time to avoid influencing the price too much, then recommend it to their customers, perhaps through their monthly house letter. When clients begin to buy the stock, it starts to go up. Like the victims who received the prescient broker's letters, other clients see the stock climbing, remember the house letter's prediction and buy into the stock. The brokerage then begins to sell the stock to the market to satisfy its own customers' demand! The brokerage actually works against its clients, many of whom will be grateful. At the top are the stragglers, the Johnny-come-latelys who miss the climb, buy in at the top, then wonder why the stock begins to go down.

Wall Street, with its legions of wildebeests, seems especially prone to what can only be called the *self-fulfilling prophecy.* Sociologists such as Richard L. Henschel at the University of Western Ontario in London, Canada, has studied the self-fulfilling prophecy in the stock market, as well as in numerous other areas of life. He cites the Tsai-Granville loop as a self-fulfilling prophesy, or SFP.

Gerald Tsai and Joseph Granville were, perhaps, the major Wall Street gurus over the last 20 years. Almost any stock that Tsai or Granville touted became so popular that the price began to rise, fulfilling the prophecy. The Tsai-Granville loop simply refers to the repeated effects of successful predictions. Starting on thinly traded stocks, which are especially sensitive to buying or selling pressures, a small-time guru can make a prediction that attracts enough attention to send the stock up when just a few believers buy in. After a number of barely noticeable successes, the miniguru catches the eye of financial journalists who need someone exciting to write about.

With increased public exposure, the guru acquires more and more disciples, not just hundreds, but thousands. What he or she says comes to pass with ever more certainty. In the loop, each success feeds the guru's credibility, which causes more speculators to climb aboard the next prediction, and so on. Scientists call this a *feedback loop.* A similar structure lies behind most self-fulfilling prophecies in other areas of life. Besides becoming famous, the market guru has an opportunity of becoming immensely wealthy. Guess who has advance information on the next stock to be touted!

The main difference between brokerages and casinos is that brokers don't seem to know they are in the gambling business. But like the casinos, the brokerages make money.

Whether the market goes up or down, as long as the trading is active, brokers make money, traders make money, investment letters make money. As long as the stock market is "important," as long as the general public believes the myths, the pattern will continue. Through it all runs a continuous, grimy thread of mathematical illusions and delusions.

Dividend Spin

I cannot resist another chart abuse gem. An alert financial journalist with the Toronto *Globe and Mail* discovered the dividend diagram shown in Figure 9 from the 1989 annual report of Bell Canada Enterprises (BCE), a large and diverse Canadian company.

The chart shows BCE dividend payments over the last five years as vertical bars. The taller the bar, the higher the pay-

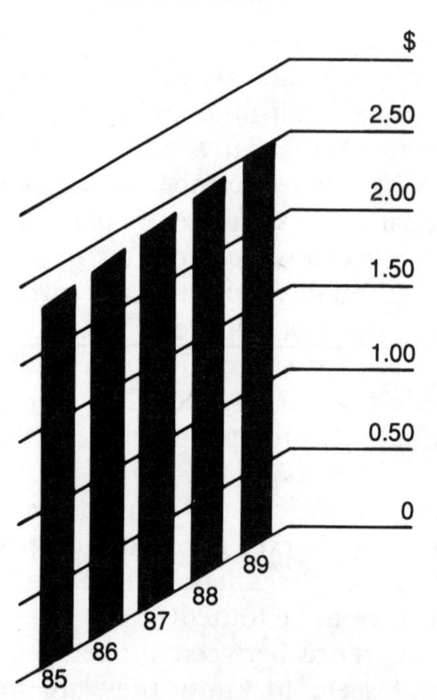

Figure 9 Tilting Chart Makes Dividend Growth Look Better

ment—you'd think. Spin doctors in the company's public relations department decided that it was okay to show any amount of upward slope in the tops of bars as long as the bottoms also sloped upward. That way, they wouldn't actually be lying. In fact, BCE's payments did not even keep pace with the cost of living over the five years in question.

The Money Forest

Money does not grow on trees, but compounded money grows even faster than a tree. In Chapter 1, I referred to compound blindness, the inability to appreciate how fast exponential growth can get. Unfortunately, in the case of money, an opposite and nearly equal decay afflicts the trees in the money forest. Decay blindness might prevent you from seeing it if you don't understand how the cost of living affects the value of money.

A statistician I shall call Mr. X writes from Sacramento, California:

> I received an offer for life insurance worth $200,000. [This much] sounds very good, but I'm 37 years old and have no plans to die before I'm 70 (even less than average once you are 37). How much would $200,000 be when I'm 70, considering an inflation of 5 percent?

Mr. X then produces a table in which the $200,000 is shown gently decaying as inflation eats it up. The $200,000 at age 37 becomes $190,476 at age 38. This, in turn, decays to $181,406 at age 39. At age 50, the purchasing power of Mr. X's life insurance has dropped to $106,064, nearly half the original value. At age 60, it has become $65,114, and at age 70 it has decayed to $39,975!

Mr. X concludes that "buying life insurance is good business if you die soon. People tend to forget that the amount they are insured for won't be the same after [even] a few years."

At the beginning of this chapter, I said that every collection of money always has a definite number connected with it—the amount. But every amount of money has a second number connected with it, the purchasing power it enjoyed a mere year ago. Decay blindness might prevent you from realizing how much the year-by-year drop in purchasing power affects the money. If inflation happens to be 5 percent in a given year, a sum of money like $200,000 will suffer deflation. The drop over time in purchasing power can be just as stunning as its nominal growth due to compounding. Banks and insurance companies, indeed any business that wants your investment money, would just as soon you remained blind to decay. Because prices inflate, money deflates.

Even those who do not suffer decay blindness might calculate deflation incorrectly. Many people (including some financial journalists), for example, think that 5 percent annual inflation will reduce the purchasing power of an amount (like $200,000) by 5 percent over a year. This is not quite true.

Suppose someone gives you $100. You could go to a supermarket and write down the prices of all your favorite brands of food until your pretend purchase comes to $100. A year later, if inflation has been 5 percent, you might expect the same cart of groceries to come to $105. The prices have increased by 5 percent. The question is, How much does that same $100 buy now and what would that basket have cost a year ago?

The short answer, developed more fully in Chapter 12, is simple. Do not subtract 5 percent of the amount from $100; divide it by 1.05. The purchasing power of the $100 drops to 100/1.05 = 95.24, a loss of roughly 4.8 percent. Not quite 5 percent, the amount can still devastate the value of a fund over time, as many senior citizens have discovered to their dismay.

The Law of Zero Return

Some people talk about the inevitability of death and taxes. The law of zero return addresses the inevitability of inflation and taxes. These two key factors play fundamental roles in a new economic law that I am about to propose. I call it the law of zero return.

The formula is no more complicated than $E = mc^2$ and is much easier to understand. In all modesty, I expect the law of zero return to eventually supercede Einstein's formula in general importance. Whereas Einstein's formula deals merely with matter and energy, the law of zero return deals with money. It goes like this:

$$\text{ROI} = \text{Taxes} + \text{Inflation}$$

It means that the return on your investment (ROI) is about equal to the taxes on your investment plus the effect of inflation on its buying power. Abuse detectives all over North America have sent me variations on this theme.

Here is just one example from a senior citizen in Richmond, California. I will call him Mr. Y since he prefers it that way. Like the millions of his fellow retirees who attempt to wring a meaningful income from their nest eggs, Mr. Y suffers from the law of zero return. He writes the following lament:

> Late in 1987 I put $10,000 into a bank certificate of deposit (CD) paying 8.75 percent, which was a fairly good rate at that time. A year later the government published the inflation rate for 1988 as 4.4 percent. It seems clear that to preserve the purchasing power of my original investment I must put $440 in with it. In other words, it now takes $10,440 to represent the [original] purchasing power of the $10,000 I invested.
>
> At about the same time, I got a 1099 form saying that I owe income taxes on the entire $875, which the government calls "profit." Our combined federal and state tax put us in the 30.8 percent tax bracket. So I paid taxes of $875 × .308, or $269.50. The arithmetic looks like this:
>
> $$\$10{,}875 - \$10{,}440 - 269.50 = \$165.50$$
>
> That is all I have to spend. But, because of the 4.4 percent inflation, it buys only 165.50/1.044, or 158.52 in terms of the money I invested a year ago. Thus the real return on my CD was 1.59 percent.

Mr. Y goes on to remark that had the inflation rate been a tad higher that year, his real return would have been exactly zero percent!

During the recession of 1991–1993, inflation dropped considerably. Does this undo the law of zero return? Not quite. Over the same period, interest rates also dropped and the law still seems to hold, albeit in an approximate way. I'm not one to speculate on the grand scheme of the economy, but the circumstances make me suspicious. It looks as if economic forces always act to preserve the law of zero return.

Very few financial planners and advisors have taken an interest in the subject of math abuse in the financial service sector. This comes as no surprise since many of them flog funds that advertise phenomenal compounding growth without mentioning inflation. Or they tout IRA schemes that make no reference to taxation. But a handful of financial advisors like Vernon K. Jacobs, who edits the *Jacobs Report* out of Prairie Village, Kansas, are prepared to advise people about precisely what happens when the law of zero return hits their retirement fund.

Blaming the government for inflation, Jacobs labels it as an additional, confiscatory tax. Jacobs has documented the twin tax ravages by applying a comprehensive analysis for the years from 1940 to 1986. He calls the law of zero return the "100 percent tax theory." A peek into the more recent past seems to confirm the law's continued operation. In a brief entitled "Why Tom Taxpayer Can't Save Enough to Be Financially Independent," Jacobs describes the theory this way:

> But, is this a new phenomenon? We have had higher income tax rates in the past and we have had higher inflation rates in the past. I set out to see if the 100 percent tax theory could be proven by actual results—by hindsight. Although I want to make some refinements in my computer analysis and the data I used, my conclusions appear to be sickeningly unescapable.

As far as Jacobs is concerned, the only way out is to concentrate on tax-deferred investments. For everyone else, the matter of making investments work for you resembles the antics of the Red Queen in *Alice in Wonderland:* It takes all the running you can do just to stay in the same place. The point is that if you don't invest, you'll just fall further behind!

6

Caveat Emptor

In the factory we make cosmetics; in the store we sell hope.
—Charles Revson, founder of Revlon

When it comes to the math abuses practiced in the commercial and business field, the time-honored Latin title of this chapter suggests the proper attitude: Let the buyer beware. Today's buyers face a battery of slick tricks and underhanded schemes that the Romans never had to deal with. Most of the abuses are on a small scale, picayune by the standards of the Wall Street financial markets or disreputable health clinics. In fact, the abuses of mathematics that lubricate business and commerce are just small enough to escape the notice of most shoppers or to be, at worst, a momentary annoyance.

But millions, hundreds of millions, of purchases are made every day and we should be aware of how many of them involve number manipulation and logical chicanery. Dare I suggest that no product line, no retail field, no business is immune to the temptations of such abuse? The amounts spent may be small but they add up, I am sure, to a discouraging figure. This chapter can only sample the myriad of math abuses you will find everywhere from car repairs to underarm deodorants.

The size of this sample of abuses is not large, to be sure, but readers will recognize many of these little tales of numerical woe: nums (numbers without the right number of digits) make it hard to figure out the contents of a package. Dramadigits (excess or arbitrary digits) make it impossible to decide whether you're paying a fair price for the product. You will certainly recognize the logical vacuum at your corner donut store, not to mention the number inflation of a dictionary publisher.

Businesses not only manipulate consumers with pricing and packaging schemes, they also band together to lobby the government in order to protect themselves against what they view as discriminatory or hostile legislation. Corporate public opinion campaigns have involved as much blatant abuse as any jiggery-pokery with prices and packaging.

Casting Out Nines

An abuse detective who recently entered a music store in Flushing, New York, looked about him, saw only nines, and suffered a common reaction. Why so many nines? Everybody knows the answer, of course: Because merchants make the prices look cheaper that way. But that's not the end of the story. A shrewd consumer should realize that when a retailer adds expenses, salaries, and overhead to raw costs, he or she doesn't always end up with prices that end in nines. In fact, fairly calculated prices ought to show every conceivable pair of digits after the decimal point. From here it is only a short step to the obvious conclusion: When retailers "change" prices to reach the magic 95- or 99-cent figure, they don't always lower them! They don't always lower a price to $19.95 from $20.18, for example. They might just raise it from $18.20.

But the abuse detective in Flushing had another complaint. People who shop intelligently want to keep track of prices in their heads, adding them as they go. The extra nines made this nearly impossible. For example, the detective purchased a record for $0.49, a tape for $1.19, and a compact disk for $9.99. The sum of the items was $11.67, the state sales tax (8.5 percent) came to $0.99, and the final price was $12.66. "There was," he writes, "no way I could do that in my head." In fact, he would gladly pay an extra penny on every purchase just for the privilege of being able to keep track of sums more easily. Where are the nums when you need them?

The abuse detective suggests rounding up all the prices by a penny and doing the sums again: record $0.50, tape $1.20, CD $10.00. Why, that comes to $11.70. If, in addition, a more easily computed (and smaller) tax were levied on the sum, coming to a final total would also be much easier on your head.

Numbers confront consumers not only from the price label but from the "contents" blurb. As any comparison shopper knows, it isn't always easy to work out what these numbers mean. A health-conscious abuse detective in Brooklyn, New York, recently examined a can of Carnation Coffee-Mate liquid, nondairy creamer. The label gave essentially the following information: serving size: 1 tbsp., calories: 16, fat: 1 gram. The detective, who wanted to limit the number of calories from fat he consumed every day to something like 30 percent (of anything he ate), merely wanted to figure out what percentage of calories from fat the can contained.

The calculation was simple enough. Equipped with one piece of extra knowledge from his calorie-counting book, that one gram of fat has nine calories, the detective divided the nine calories in the one-gram tablespoon of Coffee-Mate by 16, the total number of calories in the tablespoon. This would tell him the proportion of calories that came from fat. As a percentage, it worked out to about 56 percent, too much. Disappointed, he put the can down. But then he began to wonder: Just how accurate was the one-gram-per-serving figure? Could it be a num? What if the actual amount of fat per serving were actually more than a gram but had been rounded down by a fat-conscious company?

The lonely detective reasoned that, even if he didn't know the exact amount of fat per serving, he could at least assess for himself, right then and there, how much difference the extra fat *might* make in his calorie calculation. So he doubled the one-gram-per-serving fat content to two grams per serving. The results astounded him. By simply rounding the grams up to two, the total calorie count of the tablespoon changed to 18. This was more than the total number of calories, 16, that were there in the first place. That's when he realized how important the mystery was. For one thing, even 1.5 grams of fat would cause the calories-from-fat percentage to jump to 84 percent. If an accurate assessment of fat content is so important, why did Carnation round off the vital decimal digit? Was this an attempt to simplify numbers for consumers or a subtle scheme to num away a higher fat content than many people would accept? This one small example demonstrates a common problem that consumers have when they try to read a label. The details seem to be all there, but the numbers may have been tampered with.

Many shoppers all but ignore labels. But the same people might glance at their electric bills from time to time and wonder why they have to pay so much. Part of the reason might be innumeracy at the Electric Company—or worse. An abuse detective recently discovered some curious calculations in his gas and power bill from the San Diego Gas and Electric Company. At the top of the bill, the company showed gas consumption in "therms" as the difference between the current and previous month's meter readings. The meter had read 9,803 therms in the previous month and this month it read 9,838 therms. Most pocket calculators give the difference between 9,838 and 9,803 as 35. But San Diego Gas and Electric calculated the difference to be 36!

The detective called the company and was simply told, "No one else has complained." If no one complains, a continuing abuse of one digit can still add a powerful kick to a company's profits, as I'll explain in a minute. The central mystery of this book resurfaces in yet another case: Might the company have deliberately miscalculated the monthly differences in order to charge customers more or had it perhaps nummed the true therm readings in order to simplify things for customers? It's perfectly possible to subtract two decimal numbers, drop the decimals from both the parts and the difference, then end up with an inaccurate result. For example, if the number of therms had been 9,838.4 and 9,802.7, then the two numbers would have rounded to 9,838 and 9,803, respectively, while the original difference, 35.7, would have rounded to 36.

If this sort of thing were to happen consistently and always in a direction that favored the company, there is a tremendous "profit opportunity" as some businesspeople call it, in that one little therm. According to his bill, the abuse detective was charged about 75 cents per therm. His extra therm, multiplied across hundreds of thousands, or perhaps millions, of other customers, can obviously result in a hefty windfall for a company.

Hidden Costs

It's one thing to puzzle over the numbers you can see, but much worse when you're forced to figure out the ones you can't see. A correspondent in Rolla, Missouri, complains, for example, that a local garage has given him huge repair bills but always avoided any accusation of overcharging by citing its miniscule labor costs, barely enough to keep a mechanic in toothpicks. He suspects that the garage makes money by excessive mark-ups on the factory-ordered parts that it uses. The original costs to the garage might be considerably less than what it charges customers.

Even zealous consumers who watch costs carefully would find their numeracy of little use when faced with an invisible mark-up. The number of ways to hide costs may well exceed the capacity of a book devoted to the subject to list them. Suffice it to say that the last thing a hard-pressed shopper wants to do is go through a lengthy, difficult, and time-consuming investigation. Luckily, however, wise consumers and abuse detectives alike find their days cheered immensely by the appearance of price appeals so transparently ridiculous that they can't help laughing out loud.

Dunkin Donuts, the famous fast-food chain, recently made a strange attempt to make a price seem smaller than it really was. Fans of a new kind of muffin offered by the chain in 1991 must have been puzzled by a coupon issued by the donut maker. It showed a plate with several muffins on it, accompanied by the following irresistible offer: "FREE 3 Muffins when you buy three at the regular 1/2 dozen price."

Shoppers halfway through grade school ought to be able to sort this one out. If you buy three muffins, you get three free ones, for a total of *six* muffins. Ah, but how much do you pay for the half-dozen muffins? The half-dozen price, of course. At least the coupon doesn't mince words. It makes you shudder to think that some people thought Dunkin Donuts was offering a real deal. It makes you shudder even more to think that Dunkin Donuts thought *they* were offering a real deal!

This example, along with the next one, came from *Consumer Reports* magazine, a publication that has consistently championed shoppers' rights for many years. In a special section called "Selling It" at the back of every issue, editors relay the retail scams that readers, some of them abuse detectives, have sent in. Math abuses abound. Here is a howler from one of the 1991 issues.

A company called P.S.I. Associates found an ingenious way to make its product, a dictionary, seem better than it actually was. The *NEW Webster's POCKET PAL Dictionary* features "over 90,000 words and meanings," according to its cover. How could that many definitions fit into a dictionary that fits into the palm of your hand? Simple. The claim is literally true. The dictionary really does contain close to 90,000 words, but most of them will be found in the definitions of the 8,000 words it lists. People who thought they were buying the definitions of 90,000 words were naturally a little disappointed. The company, by the way, has no relation to the Merriam-Webster company, which must live with a name (Webster's) that belongs in the public domain.

As with the advertising schemes described in Chapter 3, consumers must beware of literal meanings. If a farmer has 20 cows and 20 bulls, you could say that the farmer has "40 cows and bulls" without being at all specific about how many of each the farmer has. And so with 90,000 "words and meanings."

It must be something of a corporate dream come true when a company charges more for a product and no one notices. According to *Consumer Reports,* it's not difficult. "Manufacturers often try to avoid a straightforward price increase for fear that consumers may not judge the product to be worth any more than they had been paying. One way to avoid the appearance of a price increase is to reduce the size of the package while holding the price steady."

Another way is to *increase* the size of the package. Here is a quick sampler of sneaky pricejacks that occurred in 1991. The makers of Mennen Speed Stick deodorant increased the size of the stick, left the price the same, and reduced the amount of actual deodorant in the stick from 2.5 ounces to 2.25 ounces. Mennen, therefore, succeeded in raising the effective price of the product by 11 percent since the stick now contains that much less product. When Fabergé's Brut, in a baldfaced scam, left price and size the same but reduced the contents from 5 ounces to 4, *Consumer Reports* moaned, "Et tu Brut?" After all, the new can said, "Now, more Brut!" When *Consumer Reports* phoned Fabergé to see how this could be possible, the company was ready. They said the new can contained "more fragrance." Oh.

Such abuses are easily discovered if you know how to figure out *unit cost,* the mainstay of all comparison shoppers. Unit cost means simply the price of an item divided by the amount of it. Sometimes you need more than division to assess a product. You might need to know a little about probability and expected value. It would be wonderful if everyone could be as alert as the husband and wife team that recently purchased a $60 portable stereo in Albany, California. The clerk wondered if they would like to buy insurance for the unit. A mere additional $20 would buy unconditional replacement for one year, no matter what part broke. The couple refused but, as the clerk rang up the sale, the manager came over and asked why they didn't want the insurance. "Even if the on/off switch breaks, you bring it in and we'll give you a completely new unit," he said.

The husband asked what percentage of the units came back every year. The manager said that around 25 percent came back. The couple thought for a moment and refused again. "I tried to explain to this guy," said the husband, "that from the numbers he had given me, the insurance wasn't worth it, all to no avail."

The couple knew that the cost of the purchase with insurance would be $80. Without the insurance, they would have to work out what statisticians might call the *expected cost,* a very simple calculation in most retail settings. Based on what the manager told them, the stereo had a 25 percent chance (or 0.25 probability) of failing. In other words, if they did not buy the insurance, there was a 0.25 probability that they would end up paying $120 for both the original stereo and its replacement. This would mean an expected cost of 120×0.25 or $30. On the other hand, there was a 75 percent chance that the stereo unit would not fail and they would end up paying only $60. In this case, their expected cost would be 60×0.75 or $45. Their total expected cost was therefore $75, the sum of these two amounts. The "real" cost of the stereo *without* insurance was $75, while the cost *with* insurance was $80. Considered as a bet, in other words, the purchase of insurance made no sense. I will explain why this is so in Chapter 12.

Putting Your Best Foot into Your Mouth

At times, companies band together in an attempt to convince the public or the government that their industry has special, hitherto unnoticed virtues, that the government has been taxing them unfairly, or that foreign competition has enjoyed an uneven playing field, and so on. Here are three quickies in the art of public persuasion. Each contains an abuse, and we have seen each abuse before.

One of our abuse detectives reports that a consortium of electric companies once mounted a public relations campaign that was intended to illustrate the positive role played by electric lighting in fighting crime. Luckily for the consortium, the detective cannot recall its name. But he made a note of the figures. The consortium's advertisement claimed that 96 percent of streets in the United States are underlit and, moreover, that 88 percent of crime takes place on underlit streets.

To most people who saw the ad, this probably looked like a strong correlation. Thank goodness for electric companies! But, wait a minute. Suppose the two percentages are exactly right. What do they say about crime and its relation to unlit streets?

Suppose you live on a well-lit street. If you had heard these figures, you might feel safe from crime. But, think for a moment. Your street is one of only 4 percent (100 – 96 = 4) that are well lit. Yet, according to those same figures, such streets enjoy a crime rate of 12 percent (100 – 88 = 12). This means that the well-lit streets have a higher crime rate than the underlit ones! To grasp the point, suppose there were 100 crimes and 100 streets. Now 88 of the crimes occurred on the 96 underlit streets. But 12 crimes occurred on 4 well-lit streets. The per-street crime rates are therefore:

Underlit Streets: 88/96 or less than one crime per street, on average

Well-Lit Streets: 12/4 or 3 crimes per street, on average

Interestingly enough, no matter how many streets or crimes you assume, you will always get a rate that is three times higher on the well-lit streets. I will use algebra in the final chapter to explain why this is so. In any event, residents of well-lit neighborhoods apparently have their local power company to thank for the higher crime rate they may be suffering.

We encountered a similar abuse earlier in the case of the Swedish life jackets in Chapter 3. There, it looked as though it might be safer not to wear life jackets unless more than 15 percent of Swedes habitually wore them. But the information of just what proportion of Swedes wore their life jackets was missing from the ad. Here, the information is complete, unfortunately for the electric power consortium. We are able to infer from the two percentages just what proportion of crimes take place on both kinds of street and we conclude that it's safer to live on an unlit street.

Another abuse seen earlier is the exaggerated threat of an extended time period, as when the figures given by the American Cancer Society scared women about the lifetime exposure risk of breast cancer as if it were some imminent figure that would take its toll in the next few months. The United States milk carton lobby, officially known as the Paperboard Packaging Council, headlined a 1988 public advertisement with the sentence:

UNIVERSITY STUDIES SHOW PAPER MILK CARTONS GIVE YOU MORE VITAMINS TO THE GALLON

According to the body of the ad, more than 20 "university studies" on the effect of light on milk conclusively demonstrated that, "in just 24 hours, milk in transparent plastic containers sitting under fluorescent lights can lose this much riboflavin (vitamin B2). . . . " A chart shows the loss of about 10 percent of riboflavin in milk after a 24-hour exposure to intense light.

You don't have to be a detective to see the real cause of concern for the Paperboard Packaging Council. Plastic containers! If the council can show something is wrong with transparent plastic containers, it will have protected the industry. Its job would have been much easier if the plastic container people had made a practice of inserting dead mice into their containers, but no such luck.

Was the ad effective? It all depended on the number of people who were able to realize that only a minuscule fraction of the light required to reduce riboflavin content in their milk by 10 percent would ever reach the containers in the dark confines of their refrigerators. After all, there are people who don't realize that when they close their refrigerator door, the light goes off.

Another abuse we saw earlier involves the use of geometric shape to distort the numerical message of a chart or diagram.

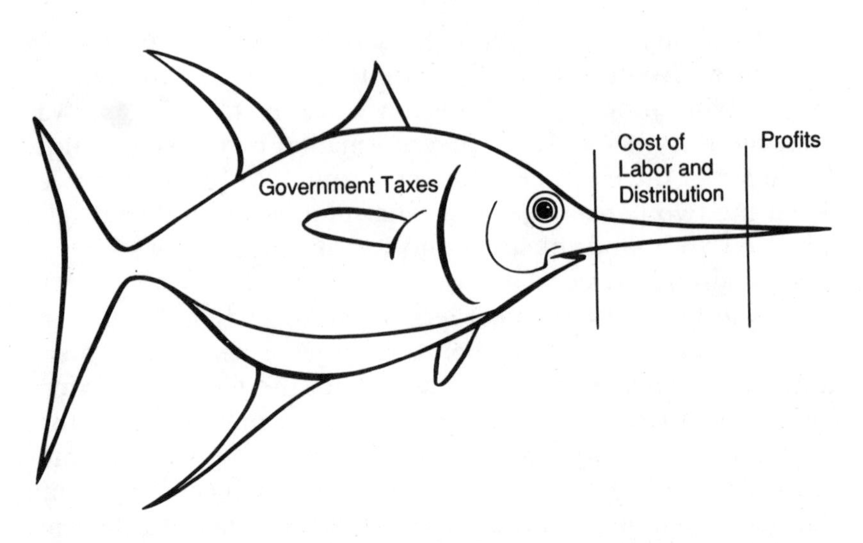

Figure 10 Fish Lobby Gives Government the Fat End of the Fish

One such abuse was perpetrated by the Canadian Association of Brewers, which lobbied its public with a creative illustration of a beer bottle used to show how many taxes the industry had to pay. This illustration was a prime example of a truly classic type of chart abuse: The shape of an object is unfairly exploited to exaggerate or minimize some factor, in this case the proportion of taxes to profits.

The approximate visual impact of the beer lobby's bottle chart can be gauged by a look at a similar instance of chart abuse by the Amalgamated Fish-Packers of America, a purely fictitious organization that zealously protects a healthy profit margin by effective constant lobbying. The vehicle of the abuse in this case is not a bottle, but a fish (see Figure 10). The profits are assigned to the long, narrow snout of the fish, while the expenses and taxes paid by the fish-packers are assigned to the fat body of the fish.

To be sure, governments have increased taxes greatly over the last few decades. Unfortunately, you will not be reassured about how governments spend the money when you read the next chapter.

7

The Government Figures

Now I think we are trying to get unemployment to go up,
and I think we are going to succeed.
—Ronald Reagan, during a 1982 Republican fund-raiser

None of us really understands what's going on
with all these numbers.
—David Stockman, Reagan's Budget Director, 1981

Democracy, a generally reasonable form of government, has one particularly worrisome drawback. People who run for public office may develop the fear of appearing brighter or better educated than the average voter. As it turns out, the fear is groundless. Too many politicians make a virtue of ignorance and a vice of knowledge.

Politicians and government officials usually abuse numbers and logic in the most elementary ways. They simply cook figures to suit their purpose, use obscure measures of economic performance, and indulge in horrendous examples of chart abuse, all in the name of disguising unpalatable truths—the level of taxation and the realities of economic performance, to name just two.

Once again, we find ourselves faced with a question of motive: Do the governments and the other public institutions described in this chapter know any better or not? As responsible organs of society, they are damned if they do and damned if they don't, rightly so in both cases. If they do know better, then they are cynically manipulating a largely innumerate public. If they don't know any better, then they must be plagued by the same innumeracy that afflicts the public. Unfortunately, the innumeracy of elected officials tends to have more serious consequences.

In perpetrating abuses, politicians set an intellectual tone that permeates government from the highest elected official to the meanest bureaucrat. By the combined weight of their sheer numbers, bureaucrats and civil servants have even more power to confuse and mislead people than any number of politicians. Needless to say, we deserve more straightforward and honest methods of presenting data, handling our money, and generally serving our needs.

Getting Elected

The whole idea of democracy revolves around a simple mathematical concept, that of majority. Is it possible for such a simple idea to become subverted by political processes? Strangely enough, when people cease to take an active interest in elections, democracy begins to evaporate. An abuse detective in Rindge, New Hampshire, recently looked into the process of elections and came away from the exercise thoroughly depressed. His reasoning went something like this.

In the 1986 U.S. election, approximately 112 million Americans were eligible to vote, about 50 percent of the population. Since 37 percent of eligible voters actually voted in that election, at most 42 million eligible voters, or about 19 percent of the population, actually cast ballots. Polling officials observe that, as a general rule, an average of something like 20 percent of the races on a ballot are not marked, even those for high offices. If the average office, therefore, received votes on only 80 percent of the ballots, this effectively reduces the number of voters in the 1986 election by another 20 percent to just under 34 million or about 15 percent.

In general, the winner of an election rarely gets more than 55 percent of the votes cast. If 15 percent of the population actually votes for a given office, even that of president, barely

8 percent of the people may have actually voted for the person who was eventually elected! And it gets worse: If you consider that half of these voters had no real enthusiasm for the candidate they voted for, thinking only to choose the lesser of two evils, you might well end up with a politician that only 4 percent of the population elected without holding its collective nose!

Some of the numbers in this innocent little excursion into percentage politics may be slightly off, but how much difference do the flaws make? If anyone can make a case for at least doubling the percentage, it would restore my faith in democracy. Meanwhile, politicians would do well to consider their real mandate.

Presidential elections are serious stuff. In recent decades, it has become fashionable for presidential candidates to debate the issues of the day on television. When Ronald Reagan debated Jimmy Carter in the autumn of 1980, the issue of public spending came up. Reagan stated that he planned to cut government spending. Carter rebutted that during his governorship of California, Reagan had increased spending by more than 30 percent. Reagan's reply, which must surely find its way into the annals of all-time numbfuscation, caught Carter totally unprepared. "Yes, but the percentage increase in the spending per capita was *less* in California during my governorship than in Georgia during your governorship."

This flabbergasted Carter and fooled most voters. Carter might at least have asked Reagan just what the "percentage increase in the spending per capita" meant, watching as Reagan's mathematical genius strayed into territory not charted by cue cards.

Per capita spending in a state refers to the average amount spent on each resident of the state. One way to calculate it is to divide the amount spent each year by the size of the population that year.

$$\frac{\text{amount spent}}{\text{population}}$$

Two things can drive per capita spending up over a given period of time: The amount spent can increase over the four-year period or population can decrease. Contrariwise, the same two factors in reverse can bring per capita spending down: Spending can decrease or population can increase.

During the years in question, the population of the State of California increased markedly while the population of Georgia stayed about the same. This difference alone was enough to put Carter's per capita spending ahead of Reagan's, a fact that had nothing to do with fiscal policy or budgetary wisdom.

Staying Elected: Cooking the Stats

Every month, the U.S. government publishes a report on the nation's economic health. Figures such as spending, income, unemployment, retail sales, and exports are reported. The figures may indicate a booming economy, one in the doldrums, or one heading downhill. Some of the figures, called *leading indicators,* are thought to forecast future economic climate. These numbers have an enormous impact on the financial markets and business generally. In particular, they influence the forecasts made by independent analysts which, in turn, affect spending decisions by businesses and governments.

A March 1992 article in *Business Week* described findings by economists Michael Waldman of Cornell University and Seonghwan Oh of UCLA that a protracted series of incorrect government reports about the economy just prior to the beginning of the 1990 recession actually made the recession worse. They traced the impact of eight consecutive leading indicator reports starting in January 1989. For example, they discovered that factory output was some $10 billion lower in the third quarter of 1990 than it would have been if estimates issued in 1989 had been correct. All low (the government later revised the estimates), the figures apparently caused up to a quarter of the shifts in economists' forecasts. In Waldman's opinion, the bad data, while not in itself sufficient to cause the recession, "clearly worsened the downturn."

As if to atone for earlier mistakes, the U.S. government later painted far too rosy a picture of an economy in grim shape. In May 1992, newspapers across North America carried a scary story of federal fudging. Ted Gibson, principal economist with the California State Department of Finance, opined that the Commerce Department had cooked stats. Gibson claimed that the 0.5 percent rise in the gross domestic product (GDP) reported by the Bush administration in the first quarter of 1992 was erroneous. Based on what Washington did with the financial data that Gibson's department sent it, Gibson inferred that all the economic estimates of performance were incorrect by a considerable margin.

For example, the U.S. Department of Commerce estimated that California nonwage income (which includes interest and private business income, among other things) was up 7 percent in 1991. In fact, it had dropped by 3 percent. The department also reported a modest decline of 1.8 percent in California retail sales when actual tax data showed a drop of 5 percent. When Commerce cited an improvement in retail sales for the first quarter of 1992, Gibson pointed to further actual declines. By February of 1991, New Jersey and other states began to notice similar distortions in how the Commerce Department reported their data, as well.

The purpose of it all? The federal government had already issued an economic forecast of a 0.5 percent gain in the GDP. According to Gibson, the Commerce Department may have been under pressure to release the cooked figures in order to conform to the national estimate. Some observers of the scene have claimed fraud and lying to Congress by the Department of Commerce. Others remark that the rosy estimates may have been crucial in a desperate campaign to reelect George Bush.

The Commerce Department certainly does not get high marks from the abuse detectives who supplied so much material for this compendium of mathematical mismanagement. How can I resist a final, fond farewell to this factory of fictional figures in the form of a graph published in 1988 in *the New York Times,* among other papers.

In a game effort to tame the federal deficit, the Commerce Department decided to divide the deficit by the gross national product (GNP). The upper graph in their published chart (see Figure 11) shows the federal deficit climbing threateningly skyward—which is what it continues to do, even today. The second graph looks much more comforting. It charts the numbers you

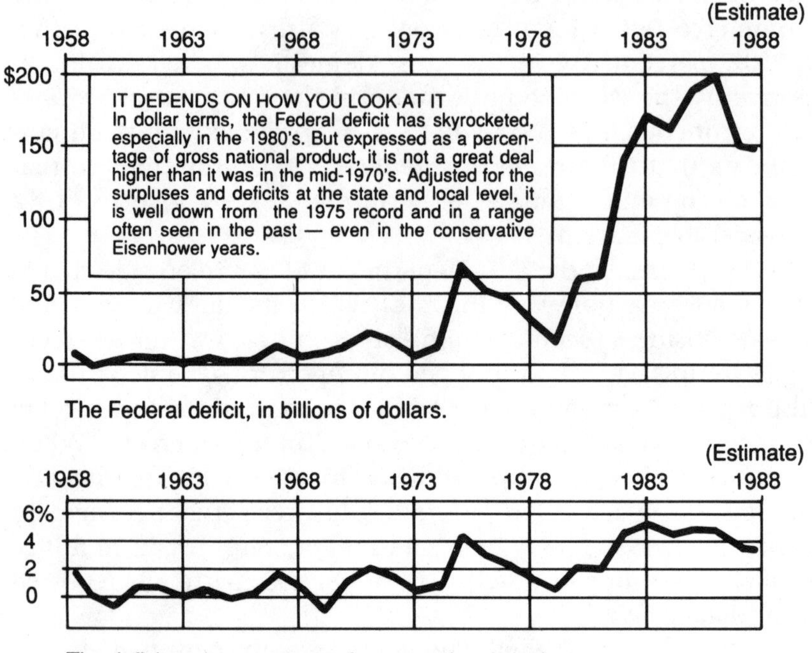

Figure 11 How the U.S. Produced a Negative GNP in 1960 and Again in 1969

get when you divide each year's deficit by the same year's GNP, expressing the result as a percentage.

The division process may have created undue strain on the Commerce Department computers. According to the lower chart, the deficit as a percentage of GNP actually went negative in certain years. What does this mean? If a division results in a negative number, then one of the two original numbers, either the divisor or the dividend, must also be negative. According to the upper chart, the deficit had never been negative in the 30-year period. This can only mean that in the years affected, the United States actually produced a negative GNP!

I will not pretend for a moment to be an economist, but it seems weird that a nation as big as the United Sates could actually produce less than nothing. The figures might be wrong, of course. It all comes back to those Commerce Department computers. They, too, may have felt the pressure to come up with numbers that would be low enough to reassure everybody. They may have divided just a little too hard.

You might ask how many people noticed the negative GNP phenomenon implicit in this curious chart. Another good question would be how many people noticed that the chart addressed the deficit and not the debt? How many people know that the *deficit* is merely the amount by which the debt increases every year? If the deficit concerns so many people, how much more worried should they be about the *debt*, which is the total amount owed by the U.S. government? Now nearly $4 trillion (see Number Numbness), the debt simply dwarfs the deficit.

Canadians Get the Picture

The Canadian government has also gotten into debt trouble. In a charming pamphlet published by the Ministry of Finance in November 1991, the Canadian government showed how it intends to get off what it called "the debt treadmill." It apparently intends to use chart abuse (see Figure 12).

The Canadian government owes more than $419 billion. While only a fraction of the U.S. national debt of nearly $4 trillion, the Canadian debt is actually larger in relation to its economy. Canada is in worse trouble than the United States.

To convince Canadians that things aren't so bad, the Ministry of Finance miraculously shrunk the national debt by cleverly squeezing it into a small black strip at the bottom of the figure. The large black block at the top shows the debt in 1984–1985; the larger white block shows how compound interest added to the debt between 1985 and 1991. Both blocks have approximately correct proportions, the larger white block of $238 billion being about 15 percent larger. After subtracting a small white bar that represents the "operating surplus," the Ministry of Finance wizards arrived at a thin black slab to represent $419 billion. By direct measurement, a block that thin cannot represent more than $49 billion. But the block is labeled "$419 billion." Its true size should be nearly that of the two upper blocks combined!

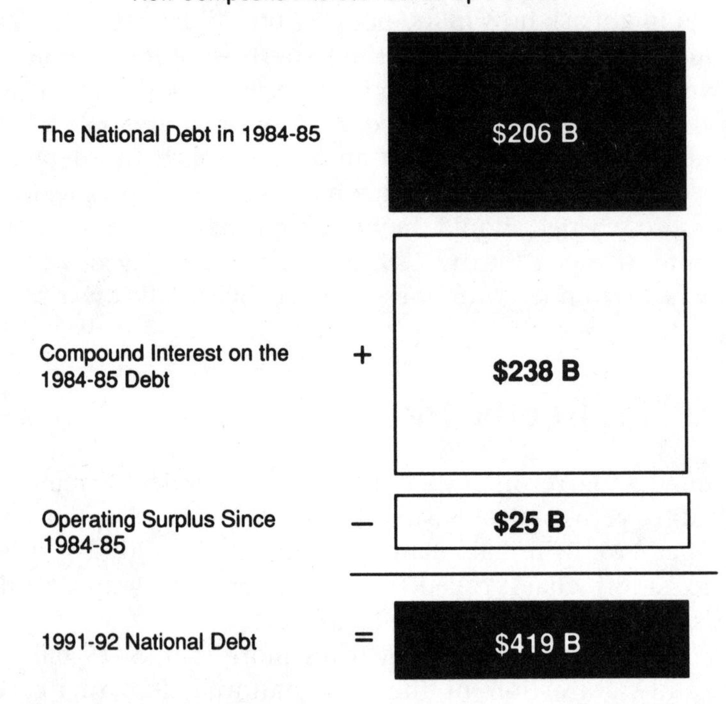

Figure 12 How the Canadian Government Tackles the Debt Problem

Here is a trick that, to my knowledge, has not yet been tried by the U.S. Department of Commerce. Of course, we will all be watching for it!

Inflated Remarks

During the 1984 election campaign, the Reagan administration trumpeted a drop in interest rates, while saying nothing about inflation. The so-called *real interest rate* means the regular interest rate corrected by inflation as described earlier in Chapter 5. When a reporter pointed out that there had not been a drop in "real interest rates," Vice President Bush reached into his deep well of humor and pointed out that "real people know what real interest rates are." This would only strike someone as humorous if they hadn't a clue what the phrase "real interest rate" actually means. Is it possible that Bush really didn't understand such a fundamental concept as real interest?

In its final, 1987 budget, the Reagan administration held funding for social programs to previous levels while increasing defense expenditures by 2 percent. Taking inflation into account, this meant that funding for social programs was actually declining. In any case, the administration forgot all about inflation when claiming that social programs had been frozen. But the Reagan brain trust suddenly remembered inflation when it came to the defense budget. The administration claimed that it had been "cut" by 1 percent. How did it arrive at this figure? It must have subtracted the 3 percent inflation rate of that year from the 2 percent budget increase in military spending.

When asked how he could account for this discrepancy, in particular the administration's failure to bring inflation into the picture in describing the social programs as "frozen," Mr. Darman, Director of the Office of Management and Budget, appealed to "common sense." He said that the person who gets $100 in 1988 sees it as the same $100 he or she got in 1987.

If you stop to think about it, you will realize that you have been hearing stories similar to these for years. Such incidents of political fudging and figure fumbling are hardly isolated. They happen, literally, all the time. We may have learned through painful experience to expect this kind of abuse by elected officials, but many people may not realize that it permeates the bureaucracy as well and even extends to government-sponsored scientific studies and weapons procurement.

More Money = Less Brains

Does the U.S. Department of Education really believe that the more money the country spends on education, the worse it gets? In early 1988, the department distributed a graph to educators showing that as education expenses rose, scores on the scholastic aptitude test (SAT) went down. Albert Shanker, president of the American Federation of Teachers, accused the government of cooking the statistics to fit its ideology that "money doesn't improve education and may even make things worse."

A quick glance at the first graph in Figure 13 seems to bear out the Department of Education's contention. Costs are steadily rising while SAT scores are steadily declining. A more careful look, however, reveals a case of range amputation. The range of SAT scores has been amputated to make its decline seem as dramatic as possible, even to the point of masking a slight increase near the end of the time period involved. The rising costs, meanwhile, are all in current dollars, meaning that inflation is not factored out of the dollar figures.

A chart corrected for inflation shows that there was, in fact, only a modest increase in educational spending between 1964 and 1986. It mostly reflects increases in the U.S. school population over the period. The declining SAT scores also look different when presented on a full-range chart. The Education Department apparently forgot to mention, in presenting its scary chart, that whereas in the past only the best students were encouraged to take the SAT, a more diverse and relatively numerous lot now takes the test.

Cooking by Radar

Many of the figures generated by governments may come from special consultants. Government departments and agencies are especially fond of hiring consultants to make decisions for them. The main charm of using a consultant lies in the responsibility factor. If something goes wrong, the department head simply says, "It's not our fault, it's those damned consultants. Believe me, we won't hire that bunch again!" It surely happens, however, that consultants are sometimes subtly directed to arrive at certain conclusions that the agency may favor.

An abuse detective who prefers not to be named worked for a period of time in the mid-1970s as a consultant in a prominent federal contract research center in the northeastern United States. A procurement agency had charged the center with the task of testing a major, new airborne surveillance system.

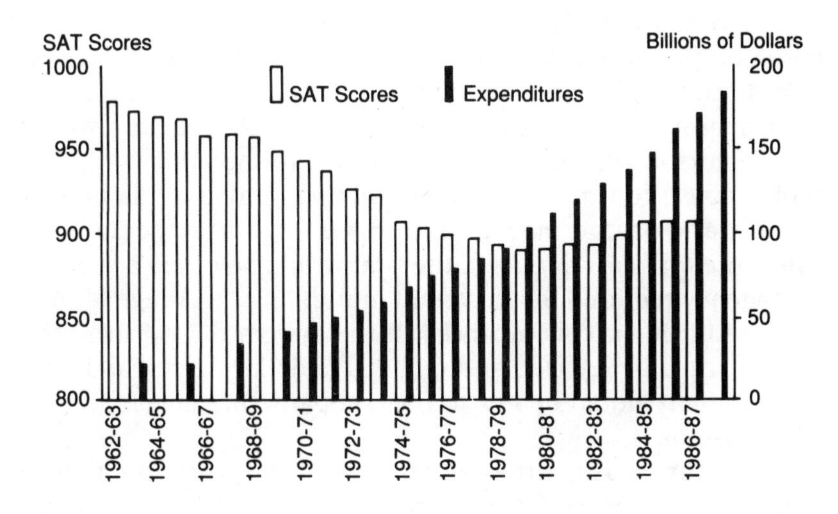

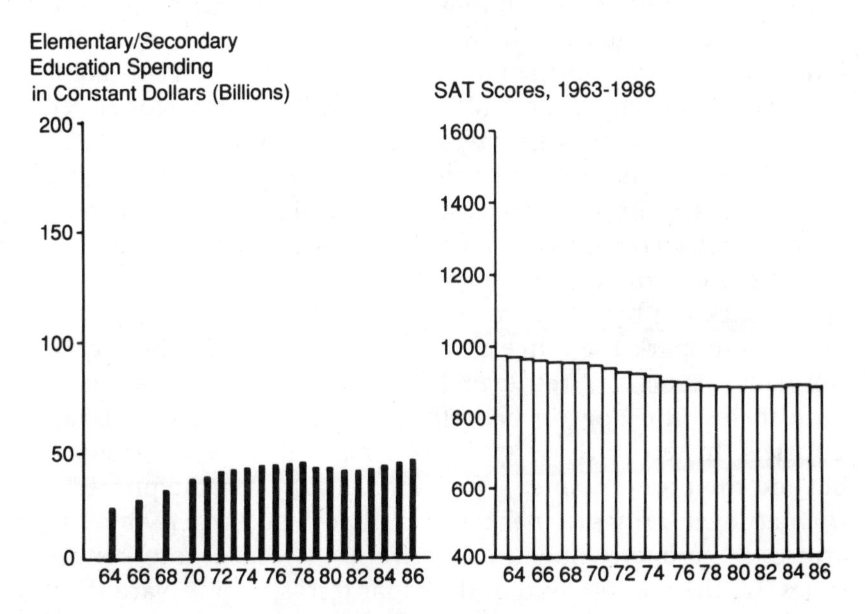

Figure 13 The Truth Behind the Education Department's Chart Abuse

Acceptance tests for new military equipment are normally based on preset, minimal levels of performance. The radar system, for example, was required to automatically track test aircraft, maintaining the track by computer. An operator would place a cursor (screen icon) on the target to show the computer which object to track. If the computer were able to maintain a stable track on the object while it was inside a special zone called the *tracking range limit*, the trial would be called a success. The agency required that a certain minimum percentage of such trials result in successful tracks. If the system did not pass the test, it would be rejected and millions of dollars in research and development would be lost to the company that developed the radar system.

The consultant witnessed an astonishing procedure that so enhanced the measured performance of the system that there was little wonder that it got accepted. Test operators decided to initiate tracks while the targets were as far away as possible, even outside the tracking range limit. If the system failed to maintain any of the latter tracks, the operators did not count them, because they were initiated outside the range limit. But if the system maintained the track once the target got inside the limit, the trial was counted a success. After all, hadn't it maintained the track inside the limit? Naturally, the contractor ended up with a performance that was "better than specification."

The test operators were, of course, guilty of introducing bias into their sample of trials. To see the bias at work, imagine testing a retriever dog you're thinking of buying by throwing a ball to the end of your property. You will buy the retriever if it brings back the ball 80 percent of the time. Your property line happens to mark the longest practical distance that you would normally expect the retriever to work within, so you intend only to test the dog on balls that stay in the yard. You run 20 trials. In ten of those trials, you happen to throw the ball beyond the property line. When the ball stays on the property, the retriever brings it back seven times, but when the ball crosses the property line, the retriever only returns it to you six times. By the criteria used in the radar test, all ten in-yard trials would be counted, whereas only six of the out-of-yard trials would enter the picture, the ones in which the retriever successfully retrieved the ball. Since the retriever returned the ball a total of 13 times, it enjoys a success rate of 13/16, or about 81 percent. The dog would pass the acceptance test even though its in-yard success rate, the one that really matters, was only 70 percent!

Perhaps the test operator was simply innumerate or suffered from an unconscious desire to see the fancy new system pass the test. But what excuse will we make for the think-tank mandarin who was out by a million?

Number Numbness in the Think Tank

An abuse detective who worked as an assistant to the President's Science Advisor in the early 1970s was given the task of farming out technology assessments to various think tanks. In their infinite wisdom, governments sometimes assess the commercial (not to mention vote-getting) potential in the development of new technologies. The assistant selected the famed Mitre Corporation to conduct an assessment of mariculture, the brave new science of marine agriculture.

When the report came back from Mitre, the assistant noticed something odd. One of the numbers in the report seemed to be a million times smaller than it should be. The assistant contacted the institution to explain his concern. Mitre recalculated, came up with the same numbers, and pronounced them to be "correct." "I don't care whether they're correct, they don't make sense," said the assistant.

A few days later, he received a visit from a senior Mitre official, a Ph.D., flanked by two technicians. The Science Advisor's assistant decided to demonstrate his problem by running his finger along one of the lines in a troublesome table. It concerned the expected bounty from various levels of fish production. They followed the line until they came to a column that carried figures for protein harvesting. The assistant remarked, "Now this is in millions." The Ph.D.'s reply was pregnant with meaning. "Oops!"

The assistant did not catch the error by applying extensive mathematical analysis or pondering deeply on the mechanics of mariculture. He simply noticed that while the fish harvest was expressed in kilogram units, the expected protein yield was given in milligrams. Simple common sense told him that kilograms (thousands of grams) of fish do not produce milligrams (thousand*ths* of grams) of protein.

Penny Larceny

Governments need money to finance their operations. The most important operation financed by governments is conducted by a special department with the word "Revenue" on the door. Abuse detectives have sent along more material on this key government function than could possibly fit here. There is math abuse in tax forms and major inconsistencies between tax laws and the way they have been implemented. But the most charming example of all is probably the most subtle. The U.S. government seems to be engaged in roundoff larceny.

All U.S. government checks to civil servants, veterans, Social Security recipients, federal retirees, and others are rounded down to zero cents. For example, the night watchman at the Federal Agency for Hangnail Control earns $2,157.72 a month, but he receives a check for only $2,157.00. This may seem a trifling matter until you realize that it amounts to a windfall of 49 cents, on average, for every check issued. The latter calculation is easy enough: If the unrounded paycheck calculation produces cents figures that range anywhere from 0 cents to 99 cents, all numbers occurring about equally often, you don't have to know all the numbers to know that they would average out to about 49 cents. After all, if you add up all the possible cents figures and divide by 100, you will get 49.5.

Suppose, for example, that the U.S. government issues 10,000,000 checks every month, a conservative estimate. The government cleans up about $5 million a month or $60 million a year. If 49 cents is penny larceny, what is $60 million?

Living With Risk

There is no safety in numbers, or in anything else.
—James Thurber, *New Yorker,* 1939

Life is full of risks. In fact, life is so full of risks that you can't really live it fully and well unless you appreciate the risks for what they are. Most people don't tremble with fear at the thought of being struck by a meteorite every time they go outdoors. Nor do they nonchalantly stroll down the middle of a busy, high-speed freeway. They understand that hefty meteorites are pretty rare and that people who wander on to freeways all too often get hit.

Other risks are not so obvious: They range from invisible carcinogens in normally safe food products to unknown mechanical flaws in apparently sound aircraft. People habitually make two errors when they imagine the risks of living: overestimating and underestimating. Two patterns of abuse and innumeracy emerge from sources that should be helping us to make these judgments. The media, more than any other group, consistently overestimates certain risks. This is probably because bigger risks sound more dramatic than smaller ones. Overestimation sells newspapers and television programs. On the other hand, some corporations that market risk-prone products, from food to flights, consistently underestimate risks.

For the numerate, risk is nothing more nor less than probability. The probability may be associated with a single event, like a bungee jump, or spread over time, like the probability of developing breast cancer (see Chapter 2).

Important social issues arise in connection with the risks associated with health and safety. We undoubtedly have a right to accurate information about the risks we live with, but assessing risks is not always easy. We may even face a risk that our measure of risk will be wrong! For example, it is possible to test negative for AIDS and yet still have the disease, and vice versa. It is also possible to test positive for drugs when, in fact, you have none in your system.

Testing the Waters

In early February 1990, thousands of cases of Perrier water were pulled from store shelves all across North America when it was reported that a quantity of benzene had accidentally gotten into a batch of the famous water in the French bottling plant. Benzene is a known carcinogen. Newspapers were somewhat vague about the risks involved, but the fact that some stores were removing Perrier from their shelves as a precaution became the main story, and what amounted to a public ban on the water ensued.

What would be the actual risk to health of drinking benzene-laced water? How do medical authorities calculate risk from a carcinogen? First, they formulate an idealized model that involves a person of a certain weight who inhales or ingests a certain amount of the carcinogen-carrying substance on a daily basis over a fixed period of time. From this risk model, to which a specific risk or probability of developing cancer is attached, authorities extrapolate by scaling in various ways. For example, a real person who has twice the weight of the idealized model person may be assigned only half the dose since there is twice as much tissue to take up the toxin. Scaling up or down from the model dosage will also alter the risk. Someone who ingests half the dose of the substance that the model does will have a reduced probability of developing cancer over the same period of time.

For example, the Environmental Protection Agency (EPA) provides cancer risk figures based on a 70-kilogram (154-pound) person consuming a carcinogen for 70 years (see Figure 14). Putting its figures together, the EPA calculates that

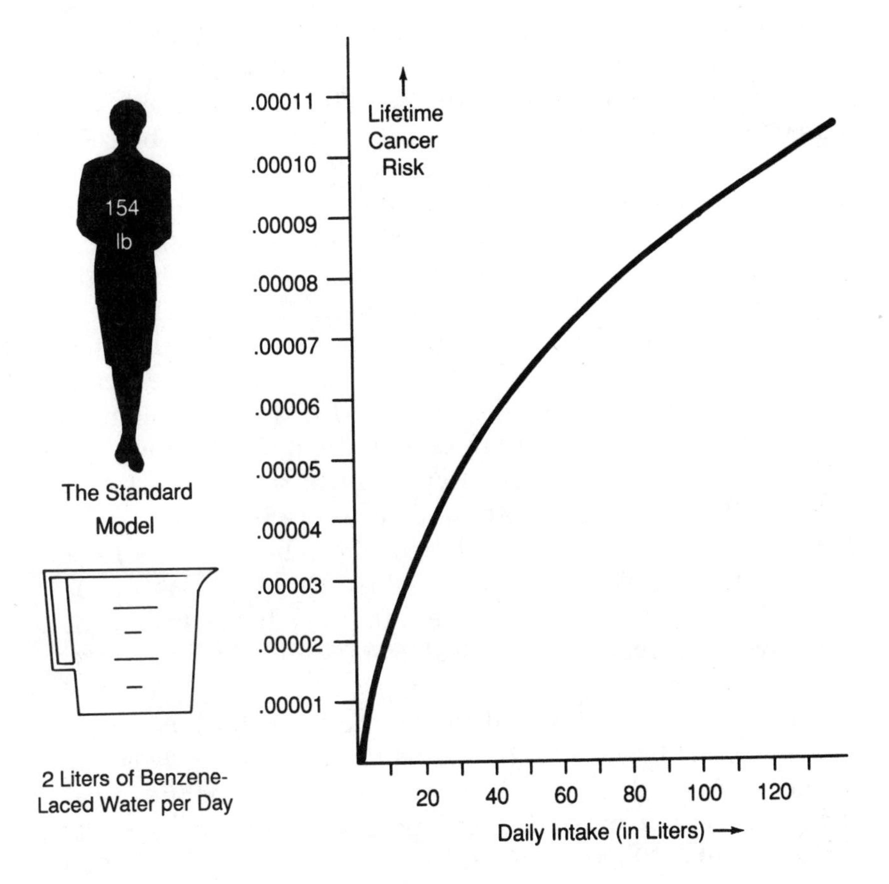

The lifetime cancer risk of drinking the contaminated water does not increase linearly (as a straight line) with daily intake. The probability of *not* developing cancer if you drink 128 liters per day is the seventh power ($2^7 = 128$) of the probability of not developing cancer if you drink 2 liters per day. The *nonrisk* trends downward with increasing velocity with the amount drunk so the *risk* trends upward with decreasing velocity, that is, more and more slowly. Not surprisingly, the model becomes less and less relevant when you explore the risk factors that result from extremes in body weight or dosage. Someone who drank 2,000 liters of contaminated water a day would develop other serious medical problems long before he or she developed cancer!

Figure 14 The Real Chances of Contracting Cancer from Benzene-Laced Water

the risk of drinking a 12 parts per billion (ppb) concentration of benzene in drinking water at the rate of two liters a day amounts to a lifetime risk of 1 in 100,000 of developing cancer.

In other words, if you weighed about 154 pounds and drank two liters of water every day at this level of benzene concentration, the probability that you would contract cancer as a direct result sometime during the rest of your life would be about 0.00001. The Perrier that found its way to North American shelves had an estimated concentration of 15 ppb, a little more than the 12 ppb used in the model. This represents an increase of 25 percent. Scaling up the probability by the same amount produces 0.000013, certainly more than the actual lifetime risk of a 154-pound person drinking two liters of tainted water every day. This represents the risk in the long run. As economist John Maynard Keynes once remarked, "In the long run we are all dead." Abuse detectives are just the sort of people who would scoff at such a small probability. Just to be on the safe side, of course, they might work out the short-run probability, say, for a year. The resulting risk figure would still be many times greater than the risk from drinking tainted water for a few weeks.

To calculate the risk of drinking the water for a year, you have to work backwards (see Chapter 12). If the dedicated drinker faces a risk of .000013 over a 70-year period, he or she obviously faces a much reduced probability over a one-year period: about 0.0000002 as it turns out. The six zeros mean that someone who drinks 2 × 365 = 730 liters of the deadly water over a year runs about the same chance of developing cancer sooner or later in his or her life as a direct result of drinking the water as he or she has of winning a 6-49 lottery after buying just three tickets. Someone who drinks the mild benzene cocktail for just a few weeks faces an even smaller lifetime risk, infinitesimal in fact.

If you think that weighing cancer risks against lottery chances amounts to comparing apples and oranges, you'd be right. But there's nothing wrong with comparing a very small apple with a very large one, especially if it helps to put things into perspective. So why not compare the risk of ultimately contracting cancer after drinking benzenated water for a year with the overall risk of developing cancer in any event? According to my trusty 1992 Houghton-Mifflin Almanac, the death rate from all forms of cancer in the year 1990 was 202.1 per 100,000. This figure can be translated directly into a probability simply by dividing the 202.1 by 100,000. The risk of someone dying from cancer in the year 1990 was, therefore, about .002. Suppose, for the moment, that this annual risk does not change much from one year to the next. Over a 70-year period, however, the total risk escalates, just as it did in the breast cancer example (see Chapter 2), into something larger, in this case, about .13. If this represents the probability of someone developing cancer sometime during a 70-year period, how much *more* should he or she worry when the risk of drinking the bad water is added to the overall lifetime risk? The figures speak for themselves:

Worry drinking untainted water: .1200000

Worry drinking tainted water: .1200002

The media, of course, rarely delve into risk for fear of discovering something that is less than astonishing. Anyway, they have other ways of dramatizing things. Enter the simple word swap. Raising a risk *by* 100 percent is not at all the same thing as raising a risk *to* 100 percent.

The Plutonium Scare

An abuse detective who lives in Victoria, British Columbia, recalls a widely reported study of plutonium contamination from weapons testing or nuclear accidents. The study estimated that 50 micrograms of the substance on the same spot in a person's lungs for 20 years would raise his or her chance of developing lung cancer by 100 percent.

Several media sources mistook the "by" for a "to" and the abuse detective found himself reading that 50 micrograms of plutonium in someone's lungs was guaranteed to kill him or her. One source went even further than this, reporting that one pound of plutonium in the atmosphere would kill everyone on earth. The overeager reporter evidently forgot (if he or she ever knew) that weapons tests alone had distributed nearly three tons of plutonium into our atmosphere. Sometimes, there are reasons to be grateful when the media get it wrong. By the same token, when the medical profession gets things wrong, there is reason to worry.

Numerically Enhanced Drugs

A large drug study described by Lynn Payer in her book *The Disease Mongers* tested the drug cholestyramine on a group of 3,806 men. The drug, which is supposed to lower cholesterol, was administered to a treatment group of 1,906 men while a placebo was given to a control group of 1,900 men.

In the first group, 30 men died of heart attacks over the 10-year study period, while in the control group, 38 men died. On the surface, it sounded as if the drug was effective. After all, it reduced the deaths by 21 percent. Shouldn't most doctors jump at the chance to use such a drug? Probably not!

Against the rather large number of men in the study, the death rates look more similar than different. The control group, for example, experienced a death rate of 2.0 percent over the period of the study while only 1.6 percent of the treatment group died. The dip from a 2 percent to a 1.6 percent death rate looks much more modest than the 21 percent reduction described in the previous paragraph. According to Payer, the real difference had "some statistical significance, perhaps, but not much practical importance." In other words, a doctor using this drug might expect to save about one man for every 250 that he or she treats with cholestyramine.

Nightlining Oat Bran

The prospect of a heart attack has so many people scared that any change in life-style that promises to stave one off will probably be embraced by millions, no matter how minimal the effects might be. How else to explain the sudden and immense popularity of oat bran? If the media were instrumental in making this innocuous cereal by-product a household word, it was only proper that the media should debunk it.

In 1990, Ted Koppel addressed the oat bran controversy on his popular ABC show, "Nightline." A researcher who had just published a debunking study in *The New England Journal of Medicine* faced off with a representative of the Quaker (Oats) Company. The researcher used his study (with a 4 percent uncertainty) to deny that oat bran had any cholesterol-lowering properties. The man from Quaker begged to differ, citing studies of his own (with uncertainties that ranged down to 2.5 percent). Even ABC's medical correspondent seemed unable to reconcile the two studies, all published in reputable journals. The debate ended in a standoff.

An abuse detective who watched the show happened to catch the uncertainties mentioned by the two debaters. The smaller uncertainties in one or two of the studies cited by Quaker owed their small size to the larger population samples used by these studies. Uncertainty, usually given as a percentage, amounts to a probability that the numbers cited in a study lie within that percentage of the correct answer. A 4 percent uncertainty, for example, may mean that the number arrived at by the study has only a small chance (usually .05) of differing from the true number by more than 4 percent.

This could mean that the debunking study had used insufficient resolution to find the effect, like someone looking for insects through out-of-focus field glasses. If it took a 2.5 percent resolution to discover any effect whatsoever, it was not only very small, but the field glasses used by the debunker were not *very far* out of focus. The Quaker representative, for his part, had nothing to gain by pointing out that the debunker's study had missed discovering an effect only because it was so exceedingly small. Widespread knowledge of this fact would certainly mitigate against Quaker's claims of a "15 percent benefit" (whatever that means).

Testing Positive

Most people feel good most of the time because most people are relatively healthy. A few people feel terrific most of the time because they have recently taken marijuana, cocaine, or something even worse.

Employers are understandably worried about the possibility that some of their workers may come to their jobs under the influence of behavior-altering drugs. In sensitive or highly responsible jobs, the outcome can be horrendous, from botched financial transactions to train derailments. Employees, however, have a certain right to privacy, especially if a proposed drug-testing program may show them to have a particular drug in their bloodstream that is not, in fact, there.

Representative Mark Cohen, a Philadelphia democrat, complained during the drug-testing debate that raged in 1987 that drug tests with even 90 percent accuracy can result in too many unfair firings. How bad can it get?

The accuracy of a drug test depends chiefly on two factors. The *sensitivity* or *true-positive rate* measures the test's ability to detect the presence of a drug at or above some minimal concentration. The *specificity* or *true-negative rate* measures its ability to detect the absence of the drug at or below that concentration. In testing employees, the sensitivity must be as high as possible to protect the employer from false-negative readings (undetected drug users). On the other hand, the specificity must be as high as possible to protect the employee from false positive readings ("detected" nonusers). Sensitivity and specificity together define the accuracy of a test.

Accuracy of interpretation is another matter. Strange as it may sound, those who are responsible for interpreting such tests must also take into account the prevalence of above-threshold drug levels in the employee's population.

Drug testing is not a black-and-white affair, anymore than is testing for AIDS. Especially germane in today's society, AIDS testing attempts to detect the presence of the HIV virus, widely considered to be responsible for AIDS. Suppose you have an HIV test done and your results come back positive. How much should you worry? And if you test negative, how much should you *not* worry? The HIV test, like most medical tests, must be interpreted statistically. Every year a handful of people who do not have the virus (or the disease) test positive. Imagine the pointlessness of their black depression. Correspondingly, imagine the false hope of those who test negative but who, in fact, have the disease!

Suppose you come from a population of 1,000 people, 10 percent of whom are known to have the HIV virus. This figure is artificially high in order to demonstrate the effect background levels can have on drug tests. A certain AIDS test is, shall we say, 98 percent accurate. It has 98 percent sensitivity and 98 percent specificity.

If you test positive, what is the chance that you actually have the disease? To find out, simply imagine what would happen if a representative population of 1,000 people were given this test. Of the 900 people free of AIDS, 2 percent or about 18 people would test falsely as positive. Meanwhile, among the 100 who actually have the disease, 98 percent or 98 people would test positive (correctly) while the remaining 2 percent, or about 2 people, would test negative (incorrectly). In other words, of the 98 + 18 = 116 people who tested positive, 18 or 15.5 percent tested wrongly.

This calculation shows us that if you came from a population in which the incidence of the HIV virus was high, such as certain central African countries, you might be found positive by a 98 percent accurate test but still enjoy a probabilistic reprieve of 15.5 percent. In other words, your actual probability of *not* having the HIV virus, even if you tested positive, would be 0.155. If you came from a population with a much lower incidence of the disease, such as North America, on the other hand, you could be more hopeful about test results that revealed the HIV virus in your system. In this case, as I will show in Chapter 12, the test has a much higher probability of being wrong, even when the true accuracy figure of 99.997 percent is used.

The background incidence of drug use, viral infection, or some other systemic agent can have a profound effect on the risk of a test going wrong. The general rule is simple: The lower the background level of the substance being tested for, the more likely that a test of a given accuracy will fail in the positive direction.

Planes (Trains) and Automobiles

Diseases of various kinds take nearly everybody sooner or later, some 94 percent of us. The rest of us die in accidents of one kind or another. Automobile accidents alone kill nearly 3 percent of us every year, airplane accidents considerably fewer. Yet people often feel more trepidation at the thought of flying 500 miles than they do of driving the same distance. Are they justified in such fears?

An abuse detective who also happens to be a reliability engineer has had the unenviable job of consulting for various major airlines, estimating the risk of accidents in their fleet of aircraft and recommending policies to reduce the probability. A favorite number used by executives finds no favor with the reliability engineer. Executives speak of the expected number of "incidents" happening in their fleet of aircraft over a period of time such as five years. The resulting number, which may be one or higher, seems more tangible to CEOs than the tiny, abstract-sounding probabilities used by reliability engineers. Yet the expected number of incidents provides a very poor basis for policy decisions such as how long to keep aircraft in the fleet before retiring them. For one thing, by this measure, you would keep aircraft longer if they happened to be in a smaller fleet because smaller fleets have a lower incidence of incidents, so to speak. Smaller fleets, after all, have fewer aircraft and, therefore, a lower overall probability of a disastrous accident over comparable periods of time.

The reliability engineer prefers *per-flight probability*. Executives may find it easier to relate to a phrase like "1.2 incidents over the next five years" than to a per-flight probability of one in a billion, but the per-flight probability allows the same policies to be adopted in fleets of different sizes. This probability measure also makes it possible to compare the safety performance of different airlines since the number of aircraft operated by an airline makes no difference to its per-flight accident probability.

Publicly, airlines like to cite the risk of flying in terms of passenger miles flown. An airline that suffered 1.2 "incidents" in five years would prefer not to advertise the fact but drown it in a larger number that means safety. The same airline, for example, might fly an enormous number of passenger miles in five years, perhaps as many as 2 *billion*. If the airline had only 1.2 incidents in that time, it could divide the passenger miles by the number of people killed in the 1.2 "incidents" and come up with a number that was still very large. Just imagine, flying 32 million passenger miles without a fatality! Creating a sense of safety that is a tad too unrealistic is not the only mischief worked by this risk measure.

According to another abuse detective, one airline advertised that it was "10 times safer" to travel by airplane than by car. Their rate of fatalities, after all, was ten times less per passenger mile than cars suffered. The same detective, however, looked at the problem in another way: If airplanes travel approximately ten times faster than cars, wouldn't their lower fatality rate per passenger *mile* translate into almost the same fatality rate per passenger *hour?* In other words, if the airline had what sounded like a super-safe rate of one fatality for every 32 million passenger miles, you might feel very safe because *you,* after all, are only traveling 800 miles, not 32 million miles.

But if automobiles have a fatality rate that is ten times higher per passenger mile, they would suffer a rate of one fatality for every 3.2 million passenger miles instead of every 32 million passenger miles, as in aircraft travel. But, at the same time, it takes a car roughly ten times longer to cover the same amount of ground as an airplane. This means that its one fatality in 3.2 million passenger miles would translate into one fatality per 80,000 passenger hours. (Simply divide the 320,000 miles by the average automobile speed of 40 miles per hour.) By the same calculation, translating the airline's figure of one fatality for every 32 million passenger miles into passenger hours results in the same figure of one fatality per 80,000 passenger hours. (This time, divide the 32 million miles by the aircraft's average speed, about 400 miles per hour.)

So what is the best way to measure the risk of flying, driving, or for that matter, taking the train? Although speed certainly *seems* to entail greater risk, this may not necessarily be the case. In the opinion of the reliability engineer, it might be better, in the case of airlines at least, to think neither of speed nor time. Since the vast majority of crashes occur either during take-offs or landings, it makes more sense to use per-flight probability. All flights, after all, must take off and land.

As far as trains and automobiles are concerned, the threat of a fatal crash or derailment seems to be more or less constant throughout the trip. In these cases, for a given speed, you may use either passenger miles or passenger hours as the measure of fatality-free travel.

Unfortunately, it takes only a headline screaming out the latest airline tragedy, hijacking, or terrorist incident to wipe out all rationality in most of us. As innumeracy writer John Allen Paulos pointed out, media reports of such incidents have seriously affected the travel industry for lengthy periods of time for no good reason at all.

9

Gee-Whiz Media Math

> The newspapers! Sir, they are the most villainous—licentious—abominable—infernal—not that I ever read them—no—I make it a rule never to look into a newspaper.
>
> —Richard Brinsley Sheridan, *The Critic*, 1779

To be fair, newspapers and other media face a difficult job: They must boil down complex, fast-moving events, a veritable stew of persons, places, and things, into a small broth that tastes fairly of the whole. That, of course, is impossible.

Reporters face two pressures when they prepare a story for publication. It must be short and it must be exciting. The first pressure may cause them to omit crucial data from their report, leaving the remaining numbers to carry a burden they will not support. The second pressure causes them to make too much of the numbers they have. In either case, we do not expect reporters to be mathematical geniuses. But we do expect them to sidestep their mind-numbing fear of mathematics long enough to ask, "Does this make sense?" "What would *I* conclude from these numbers?"

Throughout the foregoing chapters, story after story of media incompetence has been told. Though profligate, abuse of mathematics by the media does not seem to be deliberate. Instead, media machinations, hastened by deadlines, most often betray an earnest desire to impress. Think for a moment of the article in a Texas newspaper which, shortly after the fatal explosion of the *Challenger* shuttle on January 28, 1986, attempted to tote up the dollar cost of the disaster.

The reporter apparently counted the cost of building a new shuttle to replace the *Challenger*, the cost of replacing its payload satellite, an estimate of the cleanup costs, and so on. When he had nearly finished, he must have looked around for more things to add to an already impressive sum. That's when he decided to add the cost of the *Challenger* itself as well as the cost of the satellite it was carrying! The resulting figure was about twice as big as it should have been, of course.

To some people, cost is cost. Anything that might conceivably come under that category is included in a steadily growing sum. If the original shuttle cost money and a new, replacement shuttle cost money then both "costs" may cheerfully be added together to produce a new, larger "cost." Needless to say, such costing methods would not survive the scrutiny of an accountant!

This overeager desire to dramatize events through naughty numbers and illicit logic, to employ what can only be called gee-whiz math, marks the television and radio media no less than the newspapers. For example, an abuse detective sitting in his Santa Monica living room in 1990 heard a radio reporter describe the imminent fall of a satellite orbiting the earth. "The satellite's ground track takes it over many of the most populated areas in the world." The same reporter would probably be the last to add, "The satellite's ground track takes it over many *more* of the least populated areas in the world."

Sex and the Single Statistic

If you are a reporter who is eager to impress an innumerate audience, your most powerful tool will be the missing number, the absent frame of reference.

An October 1990 issue of *The Denver Post* carried the story of widespread disagreement with the *Josephson Report*, a compendium of evidence that purported to show a decline in ethics and morals among young people. A sidebar summarized some of the report's findings, including the following gem:

> PROMISCUITY: A recent published survey found 70 percent of females under 18 were sexually active in 1989,

compared with 54 percent a decade earlier; for males the increase was from 66 percent in 1979 to 72 percent last year.

Think for a moment about the phrase "70 percent of females *under 18* were sexually active in 1989. . . ." It sounds like a lot. Could the reporter have gotten it right? Imagine a roomful of sexually active females (so to speak), all under 17. Something is missing. What are their ages?

Surely the ages do not range from 17 all the way down to 1. Yet, if you assume that the females interviewed were evenly distributed across *all* age groups, a strange pattern emerges. If older females are more sexually active than younger ones, there must be a cutoff age some 70 percent of the way between 18 and 0, namely at a little over 5 years old! Below this limit, if no females are sexually active, then *all* above it would be. The reporter might as well have changed the headline (and the story) to: "100 percent of females over 5 sexually active."

The mistake lay in not specifying the age range of females interviewed, namely the lower limit. For example, if the sidebar had specified that the statistic applied only to females between the ages of 14 and 17, one could then imagine something less than 70 percent of 14-year-olds being sexually active while rather more than 70 percent of 17-year-old females are "sexually active"—whatever that means.

Science News, a normally readable and accurate science weekly, published an article entitled "US populace deemed 'sexually illiterate'" in a September 1990 issue. The article claimed that every 13 seconds a teenager in the United States acquires a sexually transmitted disease. This sounds perfectly alarming until you suddenly realize, once again, that a number is missing.

The problem does not lie with the clocklike regularity implied by the statement, even though some people take the metaphor seriously. Most people understand that the 13-second figure is a direct clue to the average incidence of such diseases among teenagers. Unfortunately, it's only a clue.

The problem is that if you don't know how many teenagers there are in the United States in the first place, you will have no way of knowing how to interpret the figure. If you calculate that for the rate to hold, about 2.5 million teenagers must be contracting sexually transmitted diseases annually. But is that a lot, relatively speaking? You'd have to look up the age distribution of the U.S. population in your almanac to find out how many teenagers there are to begin with.

Environmental Blunders

In March of 1990, the eyes of an abuse detective who had just finished reading a column on the subject of math abuse happened to light on a copy of *USA Today*. He decided to put his new insights to the test. Almost as soon as he opened the paper, he found an example.

The paper carried a report on acid rain: "More than 1,000 lakes 10 acres or larger damaged in the USA and thousands more in Canada." Again, this sounds like a lot, but how much is it? You'd have to know how many lakes of that size the United States has (see Figure 15). Worse yet, even if you knew this figure, you still wouldn't know what proportion of the total inland water volume was polluted.

To make the point more poignantly, suppose that only the five Great Lakes were affected. This represents only a tiny fraction of the total number of lakes of ten acres or more in size. Yet they represent the lion's share of the total volume. Suppose for a moment that the Great Lakes consisted of concentrated sulphuric acid while all other lakes were pure and pristine. In such a case, *USA Today* could report that acid rain really wasn't much of a problem since "only a few lakes 10 acres or larger are damaged."

George Carlin, the well-known comedian, used to perform a sketch called the "Hippie-Dippie Weatherman." In it, he would pretend to be a counterculture forecaster suffering from highs and lows of his own: "And now for some temperatures from around the nation: 58, 72, 85, 49, and 77."

For most of March and April 1986, the Saint Louis *Post-Dispatch*, serving a population of 2.5 million, published temperatures from across North America in almost as meaningless a fashion. The paper used two different unit systems without telling a soul what they were up to. An abuse detective noticed

Figure 15 Just a *Few* of the North American Lakes with an Area of 10 Acres or Larger

that the paper published all the Canadian temperatures in Celsius units and all the U.S. ones in Fahrenheit. The detective could clearly see that Buffalo, at 55, could not possibly be 43 degrees warmer than Toronto at 12, because it is just up the shore of Lake Ontario and no more than 60 miles further north.

Mildly upset over this glaring omission, the detective wrote to the newspaper to inform the responsible parties and the errors were soon corrected. Apparently, not one person had telephoned or contacted the paper in any way about the mistake, not even the people at the U.S. Weather Bureau. Why did the mistake go unnoticed for so long? The reason may lie in a common stereotype about Canada. Only Eskimos live there.

The Unexpected Expectancy

Speaking of the environment, a well-known reporter on national public radio once claimed that the pluses of modern life far outweigh the minuses. He argued that the pollution we endure is a small price to pay for the wonders and convenience of modern life. He went on to say that while life expectancy in the bad old days (before pollution) was only 35 years, now it had climbed to 72. We have never had it so good. People are living longer than ever!

What do we make of this argument? We probably all understand that *life expectancy* or *expectation* is the average age to which people live, but do we all actively imagine everyone that the average includes? The average includes not only those who live to a ripe old age, but those who die in infancy and everyone in between. Life expectancy used to be much lower in earlier times because of high infant mortality. One doubts, somehow, that those unfortunate infants just weren't getting enough pollution.

By missing the lower end of the range of ages at which humans die, the reporter had *normed* the average. As in norming probability (See Chapter 2), norming an average means failing to understand fully the total range of numbers that might go into an average. A senior citizen who also happens to be an abuse detective reports that many of his older friends look at an average like 72 and shudder, especially if they are older than that. He comforts them by showing them life expectancy tables in which those who actually reach their seventies might expect, on average, to live much longer than 72. It seems likely that the national radio reporter who normed life expectancy to rationalize pollution would probably suffer the same reaction to the number 72 when *he* reaches senior age. Everyone (even the reporter) is welcome to examine the life expectancy table in Chapter 12 to derive some small comfort in his or her statistical prospects.

Inflating Inflation

A common source of confusion that surrounds percentage figures arises from the simple notion of increase. When a percentage already represents an increase, how do we handle an increase in the increase? The following abuse has been reported in various forms, mostly in a media context.

In early October 1990, an abuse detective in West Virginia heard the anchorperson in a major network news broadcast declare that the consumer price index in September had doubled from the previous month. This sounded positively alarming. The *consumer price index* (CPI) is the cost of a standard "basket" of goods and services that is continually measured and monitored. Over time, this cost or index tends to rise and it therefore provides a standard measure of how *all* prices are changing over time. Was it possible for prices to have doubled in one month, as the reporter had claimed?

Inflation, the rate at which the price of the standard basket increases, is another matter. With inflation at 4 percent, for example, a basket of goods and services that cost $100 at the beginning of the year would cost $104 at its end. If the consumer price index were to double every month, the annual rate of inflation would have to be 537,400 percent!

Most media reports on inflation rely on a monthly inflation calculation prepared by any one of a number of responsible agencies that make such calculations their business. The normal trigger for a news bulletin on inflation is the new monthly figure. How much did the consumer price index increase last month? Perhaps it went up by 0.35 percent. Enter the sound (but to some, confusing) practice of *annualizing*.

We traditionally express all money rates, whether growth (investment) or decay (inflation), as *annual* percentages. If the monthly inflation rate of 0.35 percent continued through all 12 months of the year, it would come to 12 × 0.35 or about 4.2 percent. The calculation is not quite kosher (see Chapter 11), but it's close enough to make the point. Because monthly inflation figures fluctuate a good deal, they do not make very reliable estimators of the annual rate.

For example, an inflation figure of 0.21 percent for one month can easily become 0.42 in the next month, only to fall back to 0.32 the month after that. Could such a phenomenon have been behind the report of the doubled CPI? The detective fears it was. It was not the CPI that doubled but the apparent rate of growth of the CPI—from 0.21, say, to 0.42.

The same observer of inflated news notes that the media sometimes mistake the monthly rate for the annual one. He has heard another reporter say something like, "Inflation went up 4.3 percent last month." It didn't, of course. It stayed more or less the same. It was the CPI that went up and it went up in a way that, if continued, would result in an annual inflation rate of 4.3 percent.

Sports Reports

Abuse detectives love their sports reports, but they want their numbers and logic straight across the plate with no curve balls. They complain uniformly about one abuse that occurs across the entire spectrum of games, from basketball to hockey, the *hot streak*.

A player enjoys a "hitting streak" or a "scoring streak" during periods when he or she cannot seem to miss the net, basket, or what have you. Sports reporters encourage the notion that the player's psyche and physique have suddenly become tuned to a superhuman level of performance. This makes for increased audience tension and is, therefore, encouraged by the sports media. But suppose that hot streaks are simply statistical accidents, like the lucky streaks enjoyed by the mutual fund managers described in Chapter 5. For example, suppose a baseball player named Smith has a *batting average* of .265. This means that, over the season, he has gotten a hit 26.5 percent of the time he has stepped up to the plate. But suppose it also means that *every* time he steps up to the plate he has more or less this probability of getting a hit. Will he still have a hot streak such as 10 hits in 20 at-bats?

It is actually highly probable that such a .265 hitter will enjoy at least one such hitting streak during a season. In fact, it would be normal for him to have several. To demonstrate the point, I wrote a simple computer program that simulated a purely random .265 hitter and watched how, in the course of a "season" consisting of 600 at-bats, he enjoyed no less than three hitting streaks. Each streak, in fact, extended for 25 or more at-bats during which my random champion enjoyed a success rate of 50 percent. In short, for relatively brief periods of time (several games running) the computerized ballplayer batted .500 or better.

Numerate people can put up with "hitting streaks," allowing an amused smile to cross their features, but there is another, worse abuse that will generally set them scowling.

The air is electric as Smith steps up to the plate to face Jones, the opposing pitcher. As Smith squares off against Jones, the announcer only adds to the tension for millions of viewers at home when he says, "Smith has a lifetime batting average of .600 against Jones in night games played in domed stadiums with men on base." But how many times has Smith faced Jones with men on base during night games played in domed stadiums? Not often, you can bet. The sample of such occasions is exceedingly small, certainly too small to draw any conclusions from, yet sportscasters cheerfully spew out such useless statistics during the course of games they cover. They are aided and abetted by a small army of number crunchers who tote up every conceivable combination of conditions, poring over game records to catch the few occasions that match the ones they are likely to encounter in the coming game. In doing so, they attempt to give the network a statistical edge. The more refined the number, the more exotic the situation in which the announcer can claim some statistical knowledge, the more lively the broadcast. From such an omniscient station, sports announcers seem to have a grasp of causality itself, a grasp that certainly eludes the rest of us.

Watching his favorite baseball team on television recently, a logically astute viewer in Richmond Hill, New York, gritted his teeth when once again the announcer made what seemed a gratuitous and empty remark about the course of the game. There was a runner at first and one out when the team's best hitter came to bat. During the first pitch, the runner tried to steal second base, but was thrown out. Two pitches later, the batter swatted a home run and the announcer, quite predictably, said, "Throwing out the base runner was crucial, otherwise there would have been a two-run homer."

Here was a grave misunderstanding of causality, a refusal to grant that things might be quite different if the runner had *not* tried to steal second. For one thing, had the runner not attempted to steal second, there would still be just one man out and the batter's emotional state might have been quite different. He may have been thinking about a base hit, not driven by the anxiety that produced the home run. The pitcher, for his part, might not have thrown such a hittable ball under those circumstances, being under more pressure to strike another player out. The list is longer than this, certainly, and includes all sorts of other factors such as the wind blowing differently, the batter having an attack of indigestion, and what have you. The point is that nobody knows how things might have turned out had the runner at first *not* tried to steal second. The abuse of causality that resides in the assumption that if one thing changes in a scenario, nothing else does, is so common that the fan felt that his logic had been insulted. He suggests that the announcer would have been just as accurate to say, "They're lucky their man was caught stealing, otherwise the home run might not have been hit."

Sometimes, past events have a bearing on future ones and sometimes they don't. The law of averages, visited in Chapter 2, reappears frequently in sportscasting. Witness the complaint of an abuse detective in Webster, Texas.

The game is basketball and his favorite team's top shooter has been awarded a free throw. The sportscaster says, "Smith is a 75 percent free throw shooter and he made his previous three shots. So the law of averages says that he should miss this next shot." Whatever kind of shooter Smith is, the outcome of previous shots has no bearing whatever on the outcome of the next one, at least not for statistical reasons. Sportscasters, like many other people, seem to think that the events described by percentage statistics are obliged in some mysterious way to fulfill those statistics over the short run. Like the Las Vegas innocents who are regularly fleeced as they wait for the "law of averages" to take effect, sportscasters will suffer outcomes that run contrary to their predictions.

An informative article in the April 9, 1990, issue of *The New York Times* attempted to straighten out some of the mathematical misunderstandings that regularly infect sports reports. In particular, the author went after the notion of hot streaks. This laudable effort was marred somewhat by the miscalculation of a few key probabilities such as the 5.7 percent probability of a 10-out-of-21 hitting streak for the .265 hitter. The methods may have been correct, but somewhere a decimal slipped or a case wasn't counted. The prospect fills all of us who write on the subject of innumeracy and math abuse with horror. We must be excused, I feel, because we know the method but have simply misapplied it. Besides, we have noble motives.

The author of *The New York Times* article was turned in by an abuse detective. They are out there, waiting. Even for me.

10

The Tip of the Iceberg

Math class is tough.
—Barbie's 1992 voice chip by Mattell.

The widespread innumeracy that we find in North America today points to a failure in our educational systems to teach mathematics and the mathematical skills that matter most. Every high-school graduate should, at a very minimum, be street-proofed against the more obvious abuses described in the foregoing pages. But the innumeracy that we find in North America today hints at a far darker and deeper problem, one that threatens the existence of society as we know it.

The failure of educational systems to deliver students with even a minimal level of math and science skills has been documented with increasing frequency in recent years. The failure has its origins in two major trends in North American life. The first is a widespread ignorance of what mathematics is about and the increasingly pivotal role that it plays in the sciences, industry, and commerce. The second is a palpable decline in disciplinary standards that turns teaching mathematics (or any other subject) from a joy into a terror. Unfortunately, I can only address the first of these trends.

The Educational Crisis

As far back as 1982, the decay was already well advanced. In an international, 15-nation test survey of math skills conducted that year, U.S. students placed in the 14th rank, nudging bottom. Canadian students fared only slightly better. (Mexico did not take part in the study.) The top nation, overall, was Japan.

The results of more recent testing in the United States, as reported in *Newsweek*'s special issue on education in 1990, documented further evidence of poor math skills in students. Here is a typical question and the way students handled it:

> An army bus holds 36 soldiers. If 1,128 soldiers are being bused to their training site, how many buses are needed?

The essential step in the solution of this problem is to divide the number of soldiers, 1,128, by the capacity of the buses, 36. Amazingly, only 70 percent of the high-school students tested on the problem even thought of division or performed it properly!

$$1{,}128/36 = 31\ 1/3$$

Of these students, about two-thirds seemed content to leave the answer as it was: 31 and 1/3 buses. These students, as the *Newsweek* reporter put it, "accustomed to the sterile, self-referential world of school math courses, did not stop to question an answer involving one-third of a bus."

Any elementary school graduate should be able to get the right answer, 32 buses, without biting through a pencil. The fact that such a high proportion of high school students found the problem difficult (assuming the sample of students was a fair one) ought to alarm everybody with a stake in education.

The coming decades will involve the highest economic stress that North Americans can imagine. Lester Thurow, the well-known economist, has made some dire predictions in *Head To Head,* a book that details the changes that will have to be made in order for North America to compete successfully with the economic superpowers of Japan and the emergent United Europe. Since production facilities can (and do) migrate freely around the world, the ultimate competitive advantage enjoyed by any country or economic group will reside *solely* in its human resources. How well we do in the coming decades will depend on how well we grasp this central reality of the new economic order. As Thurow puts it,

> Firms have to be able to use new computer-based CAD-CAM technologies, employ statistical quality control, manage just-in-time inventories, and operate flexible manufacturing systems. . . . To do this requires the office, the factory, the retail store, and the repair service to have average workers with levels of education and skill that they never had to have in the past. . . . To learn what must be learned, every worker must have a level of basic mathematics that is far beyond that [currently] achieved by most American high school graduates.

Lynn Arthur Steen, editor of *On the Shoulders of Giants* and past president of the Mathematical Association of America, also worries about the educational crisis. "The typical exposure of children to school mathematics produces either confusion or, at best, a manipulation skill. But they don't seem to develop what you might call 'number sense.'" The lack of this ability, according to Steen, not only worsens innumeracy and invites math abuse, it betokens a breakdown in the ability of educational systems to supply the kind of knowledgeable human beings that a competitive economy will require.

By *number sense,* Steen means an ability to think on one's feet, to move gracefully between abstract knowledge and real situations. Even a primitive form of number sense would tell the students who failed the soldiers-and-buses question that a third of a bus made no sense. What may be lacking in the skill base of current students is a sense of mathematical immediacy, that the numbers in their heads are much more real than they think. Students also need to develop the ability to analyze in approximate terms, to be able to carry out back-of-the-envelope calculations. This skill is not only indispensable when checking calculations, but in telling us very quickly whether certain claims make sense or not.

Steen agrees with the conclusions of a report called *Everybody Counts,* put out by the National Academy of Sciences in Washington, D.C. Among the new strands the report proposes to weave into the educational fabric are a consistent instruction in data analysis and interpretation using graphs and other visual presentations, and even continuing topics in geometry, arithmetic, and algebra. The last subject, according to Steen, is just as crucial as number sense. It would develop what he calls "symbol sense" in students, the ability to use symbolic systems like algebra to solve real-world problems. The report also endorses guidelines formulated by the National Council of Teachers of Mathematics. These guidelines spell out minimal achievement levels by students in all categories by the year 2000. Although comparable to the current educational levels of other developed countries, the levels are already well above what seems to be the current standard in North America.

Given the failure of our educational system and given the looming importance of the subjects currently in decline, what can be done to change things? Simply issuing reports is not enough, according to Steen. To get the process moving, it will be necessary to develop public awareness of the issues in the same way that environmentalists have alerted the public to problems with pollution and extinction. But, given public attitudes, how do you do that?

Invasion of the Nerds

A number of popular misconceptions and stereotypes about mathematics and mathematicians threatens any public awareness program from the start. The mathematical mythology involves four distinct streams. The worst of these stereotypes focuses on a teenager who, for no apparent reason, becomes enamored of math or computers.

Nothing seems to frighten teenagers more than being thought of as a *nerd*. Always the most heavily conforming of all age groups, teenagers (and even children) have a profound fear of seeming different from their peers in any way. The popular image of the nerd with his or her glasses, pocket protector, and ready calculator (or computer) causes ordinary children to cringe as from a bug-eyed alien stepping out of a UFO. Any program that attempts to proceed without first destroying the nerd image is doomed to failure.

An adult variation of the nerd stereotype is the *boring accountant*. Never mind that accountants are not mathematicians, people often think of the two in the same way. Countless movies have portrayed the boring accountant as wimpy, obsessed with numbers and tax deductions, and relatively uninterested in human relationships, if not downright sexless. It is hardly better to be a mathematician. At parties they are all too frequently addressed as if they were accountants: "A mathematician? No kidding. Hey, I'll bet you can do your income tax in no time flat."

People also stereotype themselves in order to avoid math. Women who roll their eyes at the mention of math may be unconsciously taking refuge in the *dumb blonde* stereotype. "I was never any good at math." It may as well be Barbie speaking. If women have been culturally conditioned to avoid math, now is the time for women to break through this barrier as well. Not only is there no evidence that women have any less native ability in mathematics than men, but I will hint in the next chapter that women may have an edge over men when it comes to underlying talent. In any event, the still-prevalent stereotype of the dumb blonde continues to ensure that the vast majority of women don't even think about a mathematical career.

Finally, there is nothing so amusing as the self-stereotyping *pseudo-intellectual*, frequently but not always a person in the arts, who sniffs at the thought of mathematics. With barely a shred of mathematical education, the pseudo-intellectual not only puts mathematics down but shows pride in his or her ignorance. The funny part is that such people regularly commit strange numerical errors and fall for all the major math abuses without suspecting what is going on. It is the mathematical equivalent of someone entering a room and saying, "I been reading about this here poet called Dialin' Thomas." They may laugh if you called them innumerate but would become incensed if you ever called them illiterate.

There is no point in beating around the bush when it comes to stating the importance of mathematics today. Many scientists will be outraged to hear me say that mathematics is not only the key science, it is *the* science, *par excellence*. Eric Temple Bell, the famed writer on mathematics, called mathematics "the queen and servant of the sciences." There is no question that mathematics serves the sciences, but I also don't doubt that Mr. Bell added the "servant" part to mollify those who would feel threatened by the phrase "queen of the sciences." The simple truth is that mathematics as a field of inquiry could survive far better without other sciences than could other sciences without mathematics. A mathematician may become a scientist of any type with significantly less training; however, another type of scientist would require a great deal of training to become a mathematician. No other science matches the sharp rigor of its proofs, the breathtakingly crystalline beauty of its truths, the tremendous generality of its theorems, and the widespread applicability of its methods.

If I went on in this vein for a while in a roomful of nonmathematicians, I am sure all conversation would stop and all eyes would fall on me as on a lunatic. I could, in short, go on forever.

What Do Mathematicians Do?

Mathematicians explore abstract systems of thought. That may sound a bit formidable to some readers, but I can give a simple, nutshell example of a miniature system of thought that has all the features of bigger ones. I will give it an impressive-sounding name: *equipartition theory*. I will even make up a history for the minisubject. Like many branches of mathematics, equipartition theory began when the great mathematician François Egale was playing with a set of weights and a scale. One day he idly piled a handful of weights on one pan and then sought, among the remaining weights, ones that he could place on the other pan to balance the weights already there. It didn't take Egale long to realize that he could dispense with weights entirely and work with pure numbers.

Suppose I give you a bunch of numbers, say 6, 13, 14, 17, 31, and 55, then ask you to split them into two parts. Can you divide them up in such a way that the numbers in each part add up to the same amount? This would be like having weights of 6, 13, 14, 17, 31, and 55 grams. If you divided them up as 6, 13, 14, and 17, 31, 55 you would get a sum of 33 grams on one pan and 103 grams on the other. Hardly equal, the two numbers would not balance. You might try putting 6, 13, 14, and 17 in one pan and 31 and 55 in the other; now the two sums become 50 and 86, respectively. Closer, but still no cigar.

Insights pile up in this system of thought, at first gradually, then with increasing speed. You might already have noticed, for example, that the system does not require the numbers in each part to be in any particular order. In fact, the order isn't important at all, just *which* numbers you happen to choose for either part. That's when you notice that once you choose the numbers for the first part, you have automatically chosen the numbers for the second part by simple exclusion: If you don't choose 6 for the first part, you have automatically chosen it for the second part.

Next you notice that each part must sum to exactly half the sum you get when you add *all* the numbers:

$$6 + 13 + 14 + 17 + 31 + 55 = 136$$

In other words, the numbers in each part must add up to half of 136, or 68.

You have now simplified the problem: Find a collection of the six numbers that adds up to 68. You make a few stabs at it (mathematicians often have to guess) before you end up with 14, 6, 31, 17, and 55, 13. You have solved the problem completely and the miniature thought system called equipartition theory folds its tent and departs into the sunset of solved problems. But not quite.

What if I give you *another* bunch of numbers and ask you the same question. And what if, after that, I give you another set? And another?

The insights you have already developed amount to a theory. What applies to one collection of numbers might just apply to all of them. You already have one insight that amounts to a small theorem:

> *Theorem 1:* The sum of the numbers in each part must be half the sum of all the numbers.

The theorems help you but, after essentially guessing time after time, you begin to long for a formula, method, or technique that will find the answer for you every time. *That* is mathematics.

Your search for a method might end, ultimately, in failure or triumph. Some goals are attainable and others are not. But along the way, you would build, theorem by theorem, an edifice of knowledge based within one "miniature" thought system. Here, for example, is the next theorem you might discover:

> *Theorem 2:* If the sum of all the given numbers is odd, then no equal partition exists.

To assure yourself that the theorem really is true, you would provide a *proof,* a series of statements, each leading logically from its predecessors. The sequence leads in this way to the conclusion as stated. In the case of Theorem 2, the proof is easy. It would go something like this:

> *Proof:* Suppose a certain collection C of numbers has an odd sum but nevertheless has an equal partition. Let the numbers in the first part add up to A. Then the numbers in the second part must also add up to A since the two parts have the same sum. Therefore, the numbers in both parts add up to $2 \times A$. But $2 \times A$ is an even number, and this contradicts the assumption that the numbers in C sum to an odd amount. The contradiction means that C could not have existed in the first place.

Your common-sense mathematical mind will roam beyond this theorem, always asking natural questions. For example, if no equal partition exists when all the numbers add up to an odd value, is an equal partition *guaranteed* to exist when all the numbers add up to an even value?" Suddenly, you confront a question of pure theory. You may quickly enough discover a counterexample, 1 and 3. The question has been answered in the negative. But there is a theory, *out there,* just waiting for you. It might consist of a fast, systematic method, or algorithm, that solves any and all instances of the problem. It might consist of showing that no such method exists. It might consist of answering the general question, but only for collections of numbers having certain properties, such as being powers of two.

I have labored the point out of love. When I began to write about this particular example, I had no idea how it would turn out. Common sense led first to one theorem, then to another, then to an exciting question which I already suspected had a negative answer. It was all I could do to keep writing and not wander off into another room to ponder the matter!

There is simply no limit to how far you can go in spinning out theory, even the theory that goes with such a seemingly innocent system as our toy, equipartition theory. In this respect, mathematics resembles the other sciences. There is no end to the questions you may ask.

Junk Science

Considering the extreme closeness of mathematics to the sciences, it comes as no surprise to discover that math abuse blends imperceptibly into junk science.

For years scientists like Carl Sagan and writers like Martin Gardner have been pursuing junk science in the persons of horoscope casters, palm readers, faith healers, and various other purveyors of wishful thinking. The publication *Skeptical Inquirer* has covered these and many other abuses.

The public needs and deserves a somewhat wider appreciation of what science is and how it works, particularly with reference to some major abuses currently afoot. It is all very well for skeptical inquirers to show that the horoscope corresponds to nothing in physics, even to criticize some unfortunate developments such as *polywater* (the theory that water has structure) or cold fusion. There is much bigger game around.

What about Freud's theory of the subconscious? Most scientists understand that Freud's concepts of id, ego, complex, repression, and so on have no status as scientific terms. Recent tests of the efficacy of Freudian psychotherapy reveal no advantage over other forms of psychotherapy. Worse yet, clinical studies have shown that no form of psychotherapy in general outperforms the *placebo effect,* the well-known medical phenomenon in which a patient's expectation of getting well sometimes leads to apparent wellness.

The most damning evidence against the scientific status of Freud's theory springs from Freud's own published work, according to Frank J. Sulloway, an historian of science at the

Massachusetts Institute of Technology. Sulloway and others have revealed that Freud's own claims to having a "scientific" theory rest on just six published case histories. Not only is six far too small a sample on which to base any theory, but Freud's treatment of five of the subjects apparently had no long-term effect on relieving any of their symptoms.

How then did Freud's theories mushroom into popularity, inspiring a host of like-minded individuals to make up theories of their own? The answer lies beyond the scope of this book. Suffice it to say that those modern psychiatrists who still attempt therapies based on such shaky theories are probably misleading the public, abusing science, and, in the long run, abusing their fellow human beings.

11

Everybody Is a Mathematician

If you hang around with nice people you get nice friends,
hang around with smart people and you get smart friends,
hang around with yo-yos and you get yo-yos for friends.
It's simple mathematics.

—Sylvester Stallone in *Rocky*

Rocky, the prize fighter of movie fame, reminds a neighborhood kid that her decisions have consequences. He knows that much of life is ruled by inexorable logic. He only lacks a more detailed appreciation of how that logic and the details of life on which it operates together comprise a kind of mathematical system that everybody must navigate every day. To the extent that such a theory describes something real, we are all mathematicians. But if we are all mathematicians in some sense, why do so many of us do so poorly at math? The answer is surprising. It lies in the vast differences between the real world (where we practice our innate logical abilities) and the abstract world of mathematics itself.

The mathematics that people so readily abuse is what you might call street mathematics. This mathematics consists mostly of the simple numerical calculations of the kind described throughout this book. It also consists of what I have described (somewhat vaguely) as "logic." By this term I mean two things. First, I mean all those elements of thought that are either true or false. Second, I mean the operations that we perform on these thought elements in our heads.

As far as street math is concerned, the elements and operations of logic are both present. At one point in the shortchange transaction, the con man behaves as if the $10 he has temporarily taken from the clerk is really his. Either it is or it isn't. True or false. The shortchange artist manipulates the status of the $10 by setting up a logical framework that depends on an entirely new premise, that the $10 is actually his. The logic of the calculation (9 + 11 = 20) blends imperceptibly into the logic of the situation surrounding it. In a similar manner, distressed cancer patients are impressed by the logical framework implied by the quack clinic: the "50 incurables" they allegedly cured represent the typical outcome of their therapy. The patient may view the 50 cancer patients as a "biased sample" or, alternatively, as a "filter," but even if he or she does not have such conceptual apparatus, he or she might still question the premise and have the street smarts to ask the clinic, "Did you have any failures?" After all, anyone has the right to be suspicious of a medical facility that apparently outperforms modern medicine. Either it does or it doesn't.

In fact, the logic of street math and its calculations is hard to distinguish from a wider logic that goes by another name: common sense. When you step out of your house do you know which way the store is? You must turn either right or left. If the store is to the right, it is illogical to turn left (unless the neighbor's vicious dog brings new data into the decision). If your Aunt Mary seems angry with you, either she really is or she isn't. Every possibility branches into others. Aunt Mary may actually be angry but not at you. Or Aunt Mary may not be angry at all but was so preoccupied with another matter that she didn't even recognize you at the reception. In all cases, the explanation you give yourself is either correct or incorrect. Every time you make a choice between two alternatives, whether you are fully aware of it or not, you also make a deduction. I don't mean that you scribble formulas on a chalkboard, but rather that you weigh alternatives in your mind, trying to imagine the logical outcomes of different actions or combinations of ideas. Like mathematics itself, the world has an independent existence and does not change to suit your whims. You may wish to get to the corner store by walking only to the east but if it lies to the west, you will never get there.

When mathematicians talk about a mathematical system in its most rigorous, clear-cut form, they inevitably talk about axioms and deductions. The axioms amount to the basic assumptions or premises of the system, ideas that will not be questioned. Such "assumptions" usually encapsulate well-known properties of numbers and other mathematical objects. They aren't the sort of thing that would get a mathematician into trouble. Indeed, they are essential for the mathematical process to proceed. From these axiomatic or basic ideas, mathematicians draw conclusions in ever-widening circles of logical thought. The conclusions are all spelled out in theorems.

In life itself we never spell out theorems to ourselves but we do make logical deductions, lots of them, every day. The logic with which we operate, our common sense, depends, like mathematical logic, on certain axioms or basic ideas. By axioms, I mean not only simple operational ideas such as "if the store lies to the right, you must turn to the right" or even moral axioms such as "thou shalt not kill," but even the instinctual and compulsory drives that power our existence. Hunger, sex, and a host of other drives amount to axioms that we always obey in the absence of any logical conflict. When people talk about high-risk behaviors such as drug-taking as "illogical," they almost always mean that it is not logical for anyone who wants to live a long and productive life to take drugs. But when you recognize that drug-taking amounts to an axiom in the life of an addict, you suddenly realize that the addict's behavior is entirely logical. He or she may formulate and execute elaborate plans to obtain the money needed to finance the next fix. Given the sole axiom of the "high," its overarching importance, perhaps no one could outperform the addict in cleverness at achieving the goal. Even the behavior of so-called schizophrenics might make perfect sense if it turned out that they were merely suffering from bizarre "axioms."

The World of Games

There's a lot more mathematics in life than we realize. Let us creep up on the insight by examining more closely the places where life overlaps with mathematics as we understand it. In the end, you may agree that we are all really mathematicians of a sort.

The two chess players confront a shared world, a board of 64 squares with pieces and pawns on it. At any moment during the game, only certain moves are possible for each player, and each must conduct a logical search through the possible moves, weighing alternatives and trying to make the best one. The game involves not only the rules of play and the way that moves develop, but the two players and the way their moves interact in the shared world of the board. Because only one player can win and because both players may have a large emotional stake in the game, the purely logical dimensions of the game take on a fascinating psychological immediacy that has numerous echoes in real life. Unlike life, chess is very simple. For that very reason, we can often see many more "moves" ahead than we can in life.

Many mathematicians have noticed the mathematical nature of logical games such as chess, checkers, go, man-kala, and a host of other games that are played all over the world, some from the earliest antiquity. Such games, even with their human dimension, are *wholly* mathematical. Their mathematical structure is reflected in the mathematical *theory of games*. Far from avoiding the personal content of games, the mathematical theory embraces key elements of human behavior. It provides winning strategies for various games by analyzing not just the game but the interaction between the opposing players. It has discovered, among other things, so-called *saddle points,* game situations in which the play reaches a certain balance. Economists, sociologists, and others have picked up the theory of games and applied it to the real world. The same balance points will be found in financial auctions, deal making, negotiations, and similar situations. Here is what traders call the *saw-off,* the point where the maximum one side is willing to give up equals the minimum that the other can expect to gain.

Many conflict situations in life resemble a game—not the "games people play" of pop psychology fame, but deeper and darker contests. You may be competing for a contract, trying to outscore a conversational opponent, running against a political rival, one-upping a fashion plate, and so on. In many of these situations, each possible act has a more or less logical outcome and, if taken, will influence the course of development in a definite (if not always foreseeable) way. And in any of these gamelike situations, who doesn't think logically about the outcome?

The World of Sherlock Holmes

Another clue to the mathematical side of life comes from the detective story and its close cousin, the courtroom tale. In a detective story, there is no board, no well-defined rules of play. But the world in which the detective operates has been reduced by the author to a few characters and their actions. In these the reader must find *clues*, sometimes seemingly trivial, but nevertheless important facts that enable him or her to deduce who the culprit is. In detective fiction, there are actually two sets of opponents. First, the detective and murderer struggle together. The detective attempts to discover the murderer while the murderer works to cover his or her tracks. Second, the author and reader are also locked in a struggle. The author tries to prevent the reader from discovering the villain until the last page while continually leading him or her on. The reader, of course, wants to know from the very first page. You could even say that the detective story is also a game in which murderer and detective, author and reader, warily circle each other, probing for clues and hiding evidence.

In a detective story or courtroom drama, the evidence unfolds one step at a time, just as it does in real life. I don't mean to say that you will find out why Aunt Mary didn't speak to you by going to a courtroom or by hiring a private detective. But you will mentally review the clues you have as avidly as any reader of P. D. James. You want to *know*.

To discover the culprits in your daily life requires much sifting of evidence, bold conjectures, mistaken conclusions, and, finally, you hope, triumphant insights. Many mathematicians have noted the similarity between detective fiction and mathematics itself. Sir Arthur Conan Doyle hinted at this truth when he made Holmes's mortal enemy, the mysterious Moriarty, a mathematician!

Mathematics in Life

In a sense, we are all mathematicians—and superb ones. It makes no difference what you do. Your real forte lies in navigating the complexities of social networks, weighing passions against histories, calculating reactions, and generally managing a system of information that, when all laid out, would boggle a computer. But if this is true, why haven't you noticed this ability by now?

Anthropologists like to imagine that the human brain evolved as the result of making and using tools. Whether or not such male-oriented theories resulted from the gender of most anthropologists, it is amusing to imagine that tool use had little or nothing to do with the development of human intelligence. It makes more sense, when you think about all the complexities of human relationships and the tremendous benefits conferred by cooperation, to suppose that human evolution was driven entirely by the need to analyze relationships and predict the behavior of others. Toolmaking, considered as a mathematical activity, is utter child's play by comparison. Any fool can see where to knock the next chip off a flint spear point.

Jockeying for Position

Try explaining someone's behavior to yourself. As some behavioral scientists well know, you must imagine you are the person whose behavior you wish to explain. You must use what you know about his or her experience and inclinations. You must compare these with your own experience and draw conclusions. The fact that we actually do this points to a vast well of ability—some innate, some learned. The fact that, at the same time, we are barely conscious of the process means that we tend to overlook its importance.

To demonstrate our innate mathematical ability, consider the case of Heather and Tom. They work in the sales department of the same company. Although friends, they have occasionally competed for sales. But now they have been thrown into a truly uncomfortable situation. The position of district sales manager has just come open. Both have applied for the job but neither wants the other to know for fear of undermining their relationship. It is lunch time, and the pair are enjoying their regular Wednesday lunch at a nice restaurant.

Tom, who suspects Heather is applying, decides to probe her intentions. His goal is to get her to confess that she is applying for the job without having to confess the same thing about himself. "I hear Higgins is going to retire next month." He idly turns some lettuce over with his fork to give the impression that his mind is really elsewhere.

Heather, to whom the maneuver is transparent, suspects what Tom is up to. Her next move is designed to throw him off the track. Her general strategy, as with most men, is to get Tom confused, attack his ego in some way, and, finally, wring the confession from him without admitting an ounce of ambition. She replies, "It's true, he's going." She sighs. "You know what bugs me? When someone reaches 65, they're automatically considered incompetent. It wouldn't surprise me if Higgins fought retirement. I personally think he's good for another ten years."

Tom is surprised. Heather seems completely genuine, as though she hadn't considered applying for this very reason. Some of Tom's surprise leaks out on his face and Heather notices, smiling. That is a mistake. He sees her smile, wonders whether the statement about Higgins not retiring is a ploy, and realizes that he must demolish the nonretirement premise to get anywhere. Furiously, his mind searches through everything he knows about Higgins.

"Well, if he *does* retire, they'll probably bring in someone from the Boston office." He sighs, almost as if he had been thinking of applying for the job but had given up due to the inevitability of a Boston replacement.

"What makes you say that?"

What made him say that was the desire to read her reaction. He was aware, before he said it, that she might ask and he is ready.

"That's where Higgins came from, after all."

Now it is Heather's turn to be shocked. She didn't know that Higgins had originally come from the Boston office. For the first time, she wonders whether the position will be filled from the home office after all. What move should she make now?

The game might continue for the rest of the lunch. Move versus countermove, evidence weighed and evidence denied. Facial evidence, verbal evidence, and factual evidence all intertwined into the struggle. The brains of Tom and Heather are as busy as those of any two mathematicians ever were. Interviewed later about their discussion, Tom and Heather might deny almost everything we privileged insiders witnessed of their thoughts. For the most part, they were not aware of their thoughts as fully conscious processes.

Mathematics Is Too Simple

Suppose you grant that we are all mathematicians in some sense or that, at a very minimum, we all have this innate, largely unconscious logical ability. Why, then, do we continue to exhibit such awful innumeracy and why, for that matter, is mathematics education in such a crisis? The answer is already implicit in what I have said about mathematics in life. Mathematics itself is too simple!

By this I don't mean that the subject is simple. Far from it! No subject of human thought has anything like the stunning depth and complexity of mathematics. But the elements of mathematics, primary concepts like numbers, sets, relations, and even functions, are really quite simple. Paradoxically, it is only this simplicity (and clarity) that makes the complexity of mathematics possible.

Those of us with little or no familiarity with formal mathematics are nevertheless used to thinking complex thoughts about complex subjects, namely other people. When we come to study mathematics, we find it hard, perhaps, because we cannot get used to thinking about such simple subjects. It's much harder, for a mind that readily analyzes Aunt Mary's strange behavior at the reception, to realize that *A*, *B*, and *C* have no character or personality whatever.

In mathematical concepts, all unnecessary details have been stripped away by the process of abstraction. The naked idea stands before you and your first temptation is to clothe it with some detail, even if it means missing the whole point of the concept. Deep down you want *A*, *B*, and *C* to have human dimensions. Instead, they simply stand for numbers, sets, or some other apparently barren concept.

Humorist Stephen Leacock once wrote an amusing story called, *A, B, and C: The Human Element in Mathematics*. What makes the story so funny, I think, is that it attempts to put human character back into the algebraic names so that we are comfortable dealing with them. Although he doesn't try to bring algebraic concepts to life, he uses humans as surrogates for the confusing array of letters that the student of mathematics must face:

> The student of arithmetic who has mastered the first four rules of his art, and successfully striven with money sums and fractions, finds himself confronted by an unbroken expanse of questions known as problems. These are short stories of adventure and industry with the end omitted, and though betraying a strong family resemblance, are not without a certain element of romance.
>
> The characters in the plot of a problem are three people called *A*, *B*, and *C*. The form of the question is generally of this sort: "*A*, *B*, and *C* do a certain piece of work. *A* can do as much work in one hour as *B* in two, or *C* in four. Find how long they work at it."

This particular problem, as Leacock states it, doesn't make much sense and therefore leaves the reader with that same uncomfortable confusion experienced in grade school or high school. Leacock goes on to describe how *A*, *B*, and *C* are given a fantastic variety of jobs to perform: walking, racing, swimming, pumping out cisterns, and piling cordwood. *A* always has the best of everything. He gets the fastest locomotive, the right to swim with the current instead of against it, and so on. *B* and *C* do not so well. The characters of the three are formed accordingly:

> *A* is a full-blooded blustering fellow, of energetic temperament, hot blooded and strong willed. . . . *B* is a quiet, easy-going fellow, afraid of *A* and bullied by him, but very gentle and brotherly to little *C*, the weakling. He is quite in *A*'s power, having lost all his money in

> bets. Poor *C* is an undersized, frail man, with a plaintive face. Constant walking, digging and pumping has broken his health and ruined his nervous system. His joyless life has driven him to drink and smoke more than is good for him."

As Sheila Tobias, author of *Overcoming Math Anxiety,* has pointed out, people who are introduced to mathematical problem solving for the first time routinely try to add dimensions to the problem that simply aren't there. Used to bringing an enormous array of data to the mental table, they simply aren't ready for the utter simplicity of it all. They may attempt to fill in enough lifelike elements to make the characters real and, therefore, manageable by a mind used to more complex situations. Otherwise, the terrain is all too alien.

If mathematics is so hard for people, it may not be due to a lack of innate ability at all, but rather to a cognitive style that demands a certain level of complexity that just isn't there! If this idea holds any truth at all, it may help people learn mathematics better by inspiring them with a sense of confidence. Sometimes expectation paves the way to a completely new learning experience. If the theory holds water, it suggests that the best way to teach mathematics, at least to students encountering it for the first time, is to move gradually toward simple, abstract situations from complicated, real-life ones. Not apples and oranges at the supermarket shelf nor transactions at the cash register, but the shoppers themselves, and the logic of their social interactions.

Mathematics education could even be based, in part, on math abuse with its accompanying psychological dimensions. The next chapter contains some hints on how this might be done.

12

Street Math

> The ability to think mathematically will have to become something taken for granted as much as the ability to read a newspaper is at present. Such a change will seem fantastic to some people. But so would universal literacy have seemed absurd a few centuries ago.
>
> —W. W. Sawyer, from a speech

To achieve universal numeracy, society will have to struggle on two fronts, the classroom and the street. As I pointed out in Chapter 10, educational reforms are on the way but, like all bureaucracies, the mills of the educational institutions and authorities will grind exceeding slow. The shift to numeracy, for the present, depends on individuals who arm themselves against a thousand abuses by (a) taking extension or night courses in mathematics or, at least, (b) reading books like this one. Throughout the previous chapters, I have mentioned other books on the subject of math abuse and innumeracy. These are reviewed in the Annotated Bibliography.

In the meantime, there are some numbers left uncounted, some probabilities left dangling. Having promised to spell these out, I can also attempt to weave these explanations together into a meaningful network of ideas that will serve as a temporary snare for further abuses. It's tough out on the street. Besides knowledge of many known abuses, the ordinary citizen needs some tactics that will serve when new abuses lurch out of an alleyway. One tactic could be called "general knowledge"; the other could be called "the back of the envelope."

Strange as it may sound, abstract, general knowledge is like muscle. If you can move easily between general knowledge and specific situations, you can apply the same basic ideas to a variety of different abuses, not to mention all the problems that may come up in your work. As a general rule, you must develop the ability to abstract things on the spot, to strip away and discard all that is not numerically or logically essential in the situation. Once you have done this, the problem will almost always show itself as part of a general pattern that you already understand and that you already know how to handle. Once the problem is solved on this general level, you may bring it back down to street level, reclothing it in particulars.

A general knowledge of elementary probability theory, for example, will enable you to apply the same general idea over and over again in a myriad of instances from health risks to the fortunes of sport without ever once having to "reinvent the wheel." Even knowledge that is mildly abstract or general is better than a collection of specific recipes that apply to severely restricted cases and that prevent their practitioners from ever suspecting how easy it all really is.

As far as the second tactic is concerned, the quick and dirty calculation is often adequate for the task at hand. The "back of the envelope" serves as a metaphor. It means those calculations and deductions that you can shorten to fit the back of an envelope by making one or two simplifying assumptions that do not materially change the outcome of the calculation. How far you can simplify is a delicate judgment. It will not always be clear to you how much you can simplify things before the back of the envelope breaks down and your calculations become nonsense.

Numbers, Growth, and Decay

What concept could be simpler than numbers? For all their simplicity, numbers lead a strange existence, occurring in three distinct forms. Take the number three, for example. It exists in the real world, implicitly, in every collection of three things. When three people walk down the street together, there are the three people and there is the number three accompanying them, so to speak. At the same time, the number three exists as an abstract concept in our minds. As a "threeness," a potential threehood that might attach to any collection of three things, it awaits manifestation. Finally, the number three also lives a symbolic life as a mark on paper: 3.

How many of us appreciate what an earthshaking revolution the decimal notation brought to this planet? By simply allowing each position within a multidigit number to stand for a different power of ten, the writing of numbers took on an easy and immediate character that rendered tally sticks, knotted ropes, and even the Roman numerals obsolete. The principle is more or less obvious. A number like 5,284 amounts to an instruction to add together: 5 thousands, 2 hundreds, 8 tens, and 4 ones. Thousands, hundreds, and tens are powers of ten.

The *power of a number* means to multiply out so many copies of the number. The number of copies is the power. The notation is simple. If you just use letters to stand for numbers, you can spell out the notation once and for all:

$$N^P$$

Here, N is the number and P is the power to which it will be raised. The notation N^5, for example, means $N \times N \times N \times N \times N$. To tie up what I said earlier about our number system, its ability to express an enormous range of numbers in compact form depends, implicitly, on the power notation. After all, a number like 5,824 is really the sum of four powers, each multiplied by a digit:

$$5 \times 10^3 + 8 \times 10^2 + 2 \times 10^1 + 4 \times 10^0$$

Numeric powers play a very important role, not only within mathematics itself, but in the way mathematics describes nature. The phenomenon of compound growth is found everywhere in the natural world and within civilization itself. It occurs whenever an object or number increases its size by a fixed ratio. For example, here's a number: 123. Watch what happens when this number is multiplied over and over again by the same ratio, say, 1.2. In other words, watch what happens when 123 grows by 20 percent at each step:

123
147
177
212
255

I have omitted the decimal fraction part of the numbers for the sake of simplicity. After four applications of the ratio 1.2, the number has grown from 123 to 255, slightly more than double. If the initial number 123 represented the size of a population of buffalo or the amount of a dollar deposit in a high-yield account, the consecutive numbers might represent the size of the herd or the deposit in consecutive years. In just four years, the number has doubled. It may seem strange to some people, but the time it takes the number to double at an annual growth rate of 20 percent is independent of the size of the number. Called the *doubling time,* it depends only on the annual rate of increase.

The power of algebra is amazing. As one of the most useful forms of general knowledge, algebra can be used to prove all sorts of interesting and useful facts. In its humblest, almost trivial form, it can still demonstrate that doubling times are independent of the amount undergoing increase. If you use a marker or symbol like A to stand for the original amount, then A will undergo all the adventures that 123 did and, at the end of the symbolic calculation, you may simply measure how much A has increased. Watch now as A goes through the ratio multiplying process:

$$A$$
$$(1.2) \times A$$
$$(1.2) \times (1.2) \times A$$
$$(1.2) \times (1.2) \times (1.2) \times A$$
$$(1.2) \times (1.2) \times (1.2) \times (1.2) \times A$$

The power notation makes the final amount simpler to read:

$$(1.2)^4 \times A$$

If you multiply out four copies of 1.2, you will get 2.0736 or approximately 2.07. In other words, no matter what value A has at the start of the process, the value of the final amount will be about 2.07 times A. The doubling time for a 20 percent annual rate of increase, just a little under four years, is the same no matter what amount you start with.

Readers who want a taste of mathematical research, even in a small way, can ask the most natural question in the world: "*How* does the doubling time depend on the rate of increase?" Can you find a formula that will give you the answer in all cases, more or less immediately?

Of all common forms of compound increase, doubling can result in phenomenal growth. Once during a Toronto public radio station's drive for funds, the announcer threw the airwaves open for listeners to suggest new ways to raise money. One math prankster phoned in the following suggestion: "Why not ask people to pledge a year's worth of pennies? On the first day, they would pledge one cent, on the second day two cents, on the third day four cents, and so on. Each day they would just double the previous day's pledge until a year has gone by. Then they just mail in a check for the total."

The announcer thought this was an excellent idea and recommended it to listeners. Unfortunately, the amount of money that hapless donors must mail in at the end of the year would exceed all the pennies in Canada, all the money in the world. Even if a listener had $1,000 for every atom in the known universe, it would not even come close to the final amount owed. Yet, even if the listener had insufficient funds to cover it, he or she could still make out a check. In power notation, the amount looks deceptively small:

$$2^{365}$$

This is no reflection on the smallness of the final pledge but on the power of the power notation, so to speak.

The number just mentioned is so large it even dwarfs the NSA's googol mentioned in Chapter 1:

$$10^{100}$$

The question raised by the abuse detective after admiring the NSA advertisement was, "How fast do you have to count to reach a googol in 120 years?" For people without computers, the answer depends on using the power notation to advantage. How many seconds are there in 120 years? Most people can do this calculation without undue strain. Simply multiply the number of seconds in an hour by the number of hours in a day by the number of days in 120 years. The answer is 3,786,912,000 seconds. To find out how high you'd have to count every second to reach a googol in 120 years, simply divide the googol by 3,786,912,000. But how do you do that in your head? It seems impossible. How do you do it on a calculator when a googol won't even fit on the display register?

With the aid of the power notation, you can do the division on the back of an envelope, at least to obtain an approximate result. In power notation, after all, the problem looks trivial:

$$\frac{10^{100}}{3.8 \times 10^{9}}$$

Just by looking at the enormous discrepancy in the numbers taking part in the division, you can tell that the final result is also going to be enormous—so enormous that the exact amount you divide by isn't going to change the final conclusion at all. That's why I have taken the liberty of changing 3,786,912,000 to 3,800,000,000, or 3.8 times 10 to the ninth power. The calculation proceeds by first canceling out the powers of 10: 100 minus 9 is 91. This leaves.

$$\frac{10^{91}}{3.8}$$

The division has hardly made a dent in the googol! What will be the effect of dividing the result by 3.8? Paltry, to say the least. After all, even dividing by 10, larger than 3.8, you'd still be left with a number that had 90 zeros in it. Can you count that high in a second? Of course not.

When Numbers Combine

The marvelous decimal notation that makes numbers so easy to read and write also makes them easy to combine via the four standard operations of addition, subtraction, multiplication, and division. Have you ever tried to multiply Roman numerals? Try it on 123 × 248:

$$\begin{array}{r} \text{CXXIII} \\ \times\ \underline{\text{CCXLVIII}} \end{array}$$

We multiply numbers all the time. When computing areas and volumes of simple shapes, for example, we must multiply the separate dimensions. The connection between multiplication and areas or volumes often surprises people. The incredible expanding Toyota of Chapter 1, for example, acquired 2 additional feet when the width of the car was increased by 9 inches. The surprise in this case is that mere inches should give rise to feet. Of course it's not surprising at all when you realize that the addition of 9 inches takes place across the entire area of one side of the passenger compartment. It might be fun to run the calculation of volume backward, in effect, to see just how big the Toyota passenger compartment was to begin with. What area of the compartment would give you a 2-cubic-foot expansion when the compartment is expanded sideways by 9 inches?

The most powerful single idea in mathematics is the notion of a variable. You have already seen how the letter *A* could be made to stand for an amount of money or the number of animals in a herd. A letter or symbol can also stand for the area of the side of a passenger compartment. The great creative act of the mathematician follows: "Let *A* be the area of the compartment side." But when you multiply *A* by 9 inches to get an increase in volume of 8 cubic feet. you're in a position to know something about the Honda passenger compartment. After all, $0.75 \times A = 8$.

One of the simpler rules of algebra tells you that both sides of an equation can be multiplied or divided by the same number without violating the equality. If you divide both sides of the last equation by 0.75, you will get *A* all by itself on the left side of the equation and 10.666 on the right.

$$A = 10.666$$

In other words, you now know that the passenger compartment, if expanded sideways all over by 9 inches, must have had an area of 10 2/3 square feet. This isn't much. In order to have this area, a compartment that is only 3 feet high, for example, could not be much more than 3 feet, 6 inches long from front to back.

When one of the numbers that enters a multiplication increases, so does the product. But when one of the numbers that enters a division increases, the result can be smaller or larger, depending on what number does the increasing. This single fact leads directly or indirectly to most of the confusion that people suffer when they encounter ratios, fractions, percentages, and even probability.

How many common ratios can you name? There's the price-earnings ratio of investors, the wing aspect ratio of aircraft designers, the weight-to-power ratio of engineers, the price-performance ratio of sales campaigns, the renewal ratio of magazine fulfillment, and the capital ratio of banks. The complete list is very, very long. The usefulness of ratios lies in their ability to reduce all measurements to a common number or proportion. If two stocks have the same price-earnings ratio, for example, then both have approximately the same price in relationship to the earnings of the underlying company, even if their prices are wildly different.

$$\frac{\text{price per share of stock}}{\text{annual earnings per share}}$$

If the price-earnings ratio increases, there are two possible explanations. Either the price has gone up or the earnings have dropped. Both possibilities may even occur simultaneously.

Every ratio can be expressed as a percentage simply by multiplying the ratio by 100. Many people find it more meaningful to think in units of 100, and there are many more ratios out on the street disguised as percentages. For example, a casino's "take" may be expressed as a ratio called the *house percentage*. This is the ratio of money that enters the casino (in a multitude of pockets and handbags) that will still be in the casino when the pockets and handbags leave. Typically, it runs around 10 percent or more.

As you have seen in numerous examples of abuse involving percentages, trouble often develops when people take one percentage after another without bothering to check on intermediate results. For example, the educational report in Chapter 1 tried to imply that things were just fine now that student scores had rebounded by 70 percent after earlier dropping by 60 percent. In that chapter, I showed how a test score of 80 ended up as 55 after going through this process, dropping to 68 percent of its former value. What happens to other test scores after falling by 60 percent and increasing again by 70 percent? Will they also end up at 68 percent of their former value?

Algebraic notation and methods come to the rescue again. To use algebra, make up a symbol to denote the unknown number: *S*, for Score. What happens to *S* when you decrease it by 60 percent? *S* becomes $.4 \times S$. The new number, $.4 \times S$ now increases by 70 percent to become $1.7 \times .4 \times S$. You may multiply out the 1.7 and the .4 to get .68. In other words, no matter what value the number *S* starts with, the final result of dropping by 60 percent, then increasing by 70 percent, will always be 68 percent of *S*.

Algebra can also be used to demonstrate that in the case of the unlit streets (See Chapter 6) the crime rate is three times higher on well-lit streets. This time, the algebra will go a bit further, revealing still more of its power.

In the example of the lighting utilities ad, the claim is made that while 96 percent of American streets are underlit, 88 percent of crime takes place on them. Suppose that we assign letter names to the absolute numbers of crimes and streets involved. Suppose, in other words, that the number of streets is *S* and the number of crimes is *C*. What is the crime rate on lit versus unlit streets? According to the given percentages, $.88 \times C$ crimes occur on $.96 \times S$ unlit streets. By the same token, the opposite ratios, .12 and .04, apply to crimes on well-lit streets: $.12C$ crimes occur on $.04S$ streets. Now everything becomes simple. The two crime rates are

Lit streets: $.12C/.04S$

Unlit streets: $.88C/.96S$

The ratio of the two crime rates must be

$$\frac{.12C/.04S}{.88C/.96S}$$

The *C*s cancel out, as do the *S*s. All that's left are the numbers and when the arithmetical dust settles, you have approximately 3.3. Under the assumptions of the ad, no matter how many streets or crimes are involved, the crime rate on lit streets is *always* more than three times the rate on unlit streets!

Financial writers often add and subtract percentage changes in the value of money, seemingly unaware that this constitutes an abuse of mathematics. It is very common, for example, to read that a return of 11 percent on an investment has been reduced by 5 percent inflation to a 6 percent (real) return. In fact, this isn't quite how inflation works.

If prices rise by 5 percent, then the value of money drops by a corresponding amount, not 5 percent but 4.8 percent. For every level of price inflation there is a corresponding level of money deflation. You can make up your own formula to show the relationship between price inflation and money deflation by assigning a symbol, F, to mean price inflation. In algebraic work, it is always better to use ratios directly and not percentages, so F means the ratio by which prices rise. A 5 percent inflation figure will correspond to a value for F of .05. The question is, how does an inflation ratio of F affect the purchasing power of money? How much less, in other words, does a dollar buy this year than it bought last year?

If a basket of widgets cost D dollars last year, it would cost $D \times (1 + F)$ dollars this year after an annual inflation of F. This means that every dollar last year would purchase $1/D$ widgets. This year, that same dollar purchases only $1/D(1 + F)$ widgets, and that's all there is to it. The ratio of the second figure to the first gives you deflation expressed as a ratio. The *D*s cancel out.

$$\text{Deflation} = 1/(1 + F)$$

Once you have a working formula, you can play with it to extract all kinds of valuable information. For example, you can try out different rates of inflation on the formula to see what sorts of money deflation they give rise to. Try 10 percent. Here, the deflation turns out to be 1/1.1 or 9.1 percent. The larger the price inflation, the more widely separated it becomes from its corresponding money deflation.

Many ratios must lie between 0 and 1 in order to make any sense. Probability is one of these.

Working with Probability

What is your chance of winning if you buy a ticket in the 6-49 lottery? An important rule to use whenever you get stuck using mathematical formulas or ideas is to go back to the definition. *Probability,* as I defined it in Chapter 2, is actually a simple ratio or fraction. Here is another general idea or pattern that you may and must apply almost every time the subject of probability comes up:

$$\frac{\text{number of favorable outcomes}}{\text{total number of possible outcomes}}$$

If you buy just one ticket, the "number of favorable outcomes" must be one since of all the numbers that might come up, only the number on your ticket could be regarded as favorable. As far as the "total number of possible outcomes" is concerned, you have to know how many six-number combinations there are of numbers between 1 and 49.

A branch of mathematics that many people find especially accessible has been the helpmate of probability and statistics from the earliest days. Called *combinatorics,* it has several missions, the chief of which is counting all manner of sets, combinations, relations, networks, and a huge variety of mathematical objects that go by the name of discrete structures. The number of six-number combinations can be found with the aid of a multiplier rule. Think about the six-number combination for a moment. How many ways could the first number be chosen? Since there are 49 numbers to choose from there are 49 ways the first number could be chosen. But the second number in the combination cannot be the same as the first number so the second number can only be chosen in 48 ways. The multiplier rule enters the picture at this point: For *each* of the 49 ways of choosing the first number there are 48 ways of choosing the second one. In other words, the number of ways of choosing the first two numbers is 49×48.

The calculation doesn't end there, of course. For each of the 49×48 ways of choosing the first two numbers there are 47 ways of choosing the third number. This means there are $49 \times 48 \times 47$ ways of choosing the first three numbers. If you want to leap ahead at this point and write out the total number of six-letter combinations as $49 \times 48 \times 47 \times 46 \times 45 \times 44$, you'd be right. Unfortunately, the resulting product overcounts the total number of combinations because it treats different orderings of the same six-number combinations as being different. It would count 19, 3, 42, 8, 33, 12 as different from 8, 19, 12, 42, 3, 33. But as far as lottery tickets are concerned, these are considered to be the same combination. In fact, every combination is counted the same number of times. To repair the damage, we have to know how many times and then divide the product above by this amount. Now we find ourselves faced with a different question. How many ways can we order six different numbers?

The multiplier rule enters the picture once again. There are six ways to choose the first number, five ways to choose the second, four ways to choose the third and, altogether, $6 \times 5 \times 4 \times 3 \times 2 \times 1$ ways to order six numbers. To divide the product above by this product, all you need is a calculator. The first product comes to 10,068,347,520. The second product comes to 720. The final fraction works out to 13,983,816.

As far as the probability of your winning that lottery is concerned, you have only one chance in 13,983,816 of winning. If you divide 1 by this large number on your calculator, you may get a strange answer like 7.1511E-8. This means that the poor machine's registers have been exceeded and it has resorted to power notation. The "E" separates the number to be raised to a power from the power itself, in this case –8. The number 7.1511 is divided by 10 raised to the 8th power. Just move the decimal point eight places to the left:

$$\text{probability} = 0.000000071511$$

The additional digits past the 7 are pretty pointless, so you can omit them to find exactly the number I gave in Chapter 2 as the probability of winning the 6-49 lottery.

In the lottery example, I also used a back-of-the-envelope calculation to find out the probability of being killed by lightning sometime in the next year. This comparison bothers some people because they think the probability does not apply to them. "I'm not the type to go wandering in the woods," they might say. That could be perfectly true. People who spend less time out of doors in general are less likely to be struck by lightning. This serves to introduce two important notions called *a priori* and *a posteriori* probability. The first term means "before the fact" and the second term means "after the fact." If you know nothing at all about a person, you could say that his *a priori* chance of being killed by lightning in the next year is one in a million. But if you know that he or she is confined to a wheelchair and rarely gets out into the country, you'd be justified in calculating a new probability by attempting to discover how many people in wheelchairs are killed by lightning every year, then dividing the amount by the total number of people in wheelchairs. Such a probability, if you could ever get your hands on reasonable estimates for these numbers, would be called the *a posteriori* probability.

When calculating the probability of an event spread out over time, you may be forced to use the multiplier rule for compound probability. Let's take another look at the cruel game of Russian Roulette. There is but one bullet in the revolver's six chambers and when you spin them, point the gun at your head, and pull the trigger, there is a 1/6 chance that you will shoot yourself. The *complementary probability*, that you won't shoot yourself, is therefore 5/6. Since you need to know the probability of surviving six rounds of the game, you must find out what the probability is of *not shooting yourself* six times in a row. The event of not killing yourself on six consecutive spins is called a *compound event* and its associated probability is called a *compound probability*. Such a probability requires the *multiplier rule*. Multiply the separate probabilities of not being shot (in a single round of the game) together, one for each round. The probability of surviving six rounds is

$$(5/6)^6$$

The number works out to .335. Since this is the probability of surviving, the probability of not surviving is the complementary probability, 1 – .335 or .665.

If the abuse examples in foregoing chapters are any indication, the calculation of compound probabilities permeates street math. You must use the method, for example, in calculating how the probability of a daily telephone coincidence translates into a monthly one. In Chapter 2, I constructed a model situation in which the probability of someone telephoning you on the very minute after you thought of him or her was 0.0001. This sounds like a pretty small chance, but over a period of 10 years the probability of such an event certainly climbs. In a 10-year period with more than 3,650 days in it and with two phone calls a day, there are more than 7,300 opportunities for such a call to take place. As before, simply turn the probabilities around and ask, "What is the chance that such a call *won't* occur?"

With one call, the probability is 0.9999. Most calculators with a power-raising function will cheerfully register to the 7,300th power. This represents the probability that such a call will not occur in a 10-year period. Switching back, finally, the probability that such a call *will* occur is 0.5181 or approximately 0.52, certainly a better than even chance.

This demonstration is enhanced somewhat when you realize that thinking about someone may well continue for more than a second or two. There's even a good chance you will still be thinking about him or her less than a minute later when he or she calls you!

Sometimes you have to apply the multiplication rule in reverse gear. Suppose you have to work backward, from a probability that applies to a long period of time to a probability over the short term. According to the American Cancer Society in Chapter 2, the lifetime *(a priori)* probability of a woman developing breast cancer is about .11. Over the much shorter term of a single year, a woman under 50 faces a greatly reduced risk of .001 that she will develop breast cancer. How can you work backward from a long-term probability to get one for the short term? I will illustrate the method by using the example of benzene-laced water.

According to the standard model developed in Chapter 8, a 154-pound person drinking two liters (a little over two quarts) of benzene-laced water a day faces a lifetime risk of no more than .000013. If this represents the risk over a 70-year period, what is the risk over a 1-year period? To find that risk, you must work with nonrisk and complementary probabilities. The chance of not developing cancer over a lifetime (70-year period) of drinking the contaminated water is 1 – .000013 or .999987. Let's be algebraic and suppose that the chance of *not* developing cancer as a result of drinking the water in any one year is *P*. Now all you have to do is solve an equation for *P*:

$$P^{70} = .999987$$

What number, when you raise it to the 70th power, will give you .999987? There are various ways of getting the answer. For example, you could take the 70th root of both sides of the equation. The left side would become *P* and the right side would become the 70th root of .999987. A good quality pocket calculator or a computer (if necessry) will tell you that the 70th root of .999987 is

$$P = .999999814$$

The complementary probability, approximately .0000002, represents the chance that the dedicated drinker of bottled water will develop cancer as a result of his or her thirst in any one year. In fact, since this assumes that the connoisseur will drink the bad water for 70 years and not just for 1 year, the actual probability will be much lower than this, perhaps about the same as being killed by lightning right after buying a lottery ticket. Such morbid thoughts bring mortality tables to mind.

Morbidity, Mortality, and Worse

The difficulty experienced by seniors over life expectancy reflects our tendency to norm statistics, to expect events to reflect the average exactly instead of demonstrating the variation that always underlies such one-number summaries. In the case of life expectancy, the best medicine is a dose of the mortality table. The figures below were compiled on the basis of 1988 data from the U.S. Department of Health and Human Services (as reported in my trusty 1992 Houghton-Mifflin Almanac).

Age	*Expectation*	*Mortality*	*Age*	*Expectation*	*Mortality*
0	74.9	10/1000	31	45.9	1.4/1000
1	74.7	0.7/1000	32	45.0	1.5/1000
2	73.8	0.5/1000	33	44.1	1.5/1000
3	72.8	0.4/1000	34	43.1	1.6/1000
4	71.8	0.3/1000	35	42.4	1.7/1000
5	70.8	0.3/1000	36	41.3	1.8/1000
6	69.9	0.3/1000	37	40.4	1.9/1000
7	68.9	0.2/1000	38	39.4	2.0/1000
8	67.9	0.2/1000	39	38.5	2.1/1000
9	66.9	0.2/1000	40	37.6	2.2/1000
10	65.9	0.2/1000	41	36.7	2.3/1000
11	64.9	0.2/1000	42	35.8	2.5/1000
12	64.0	0.2/1000	43	34.9	3.0/1000
13	63.0	0.3/1000	44	33.9	3.0/1000
14	62.0	0.5/1000	45	33.0	3.1/1000
15	61.0	0.6/1000	46	32.1	3.4/1000
16	60.1	0.8/1000	47	31.3	3.8/1000
17	59.1	0.9/1000	48	30.4	4.1/1000
18	58.2	1.0/1000	49	29.5	4.5/1000
19	57.2	1.0/1000	50	28.6	5.0/1000
20	56.3	1.1/1000	51	27.8	5.5/1000
21	55.3	1.1/1000	52	26.9	6.0/1000
22	54.4	1.2/1000	53	26.1	6.6/1000
23	53.5	1.2/1000	54	25.3	7.3/1000
24	52.5	1.2/1000	55	24.4	8.0/1000
25	51.6	1.2/1000	56	23.6	8.7/1000
26	50.6	1.2/1000	57	22.8	9.6/1000
27	49.7	1.2/1000	58	22.0	10.5/1000
28	48.8	1.2/1000	59	21.3	11.5/1000
29	47.8	1.3/1000	60	20.5	12.6/1000
30	46.9	1.3/1000	61	19.8	13.8/1000

Age	Expectation	Mortality	Age	Expectation	Mortality
62	19.0	15.0/1000	74	11.2	39.2/1000
63	18.3	16.2/1000	75	10.7	42.4/1000
64	17.6	17.4/1000	76	10.1	46.1/1000
65	16.9	18.7/1000	77	9.6	50.1/1000
66	16.2	20.1/1000	78	9.0	54.6/1000
67	15.6	21.8/1000	79	8.5	59.7/1000
68	14.9	23.7/1000	80	8.1	65.5/1000
69	14.2	25.8/1000	81	7.6	72.1/1000
70	13.6	28.2/1000	82	7.1	79.7/1000
71	13.0	30.7/1000	83	6.7	88.5/1000
72	12.4	33.4/1000	84	6.3	98.8/1000
73	11.8	36.2/1000	85	6.0	n/a

Beside each age there are two figures. The first, called *expectation,* indicates how much longer the average person who reaches that age can expect to live. The second figure expresses *mortality per thousand.* For each age, it gives the probability that the person who reaches it will not live to see another birthday. The numbers embrace males and females of all races. As such, they represent a composite picture of *a priori* data. A breakdown of the same figures by race or sex reveals different numbers with generally higher expectations among whites than other races and higher expectations for females than males.

To read the table, let's suppose that you're 37. Your life expectancy, 40.4, means that you can "expect" to live another 40.4 years and reach the ripe age of 37 + 40.4 = 77.4. That may sound wonderful but, wait, there's mortality. The mortality figure for your age group is 1.9 per thousand, which means that two out of every thousand of you (37-year-olds) will die in the next year. Those people will obviously not live to see 77. If you reach 77, on the other hand, you are not automatically dead any more than you automatically lived beyond 37. At age 77, the expectation has dropped to 9.6 years and about 50 of you per thousand will not live beyond 77.

At such a point there is always a wiseacre who thinks, "Gosh, I could live forever. Let's see: 77 + 9.6 = 86.6. All I have to do is look up 86 in the table and. . . ."

The table, unfortunately, does not extend beyond 85 as the statistics become too sparse to be reliable at this point. But the eternity game is not playable, in any event. At some point, the expectation effectively drops to zero and you are living on borrowed time, so to speak.

Probabilities that depend on shifting underlying distributions may have to be revised from time to time. Take the case of the AIDS test mentioned in Chapter 8. When the background population suffers a high incidence of AIDS, a test becomes less likely to produce a false positive. In the population mentioned, 10 percent had AIDS and those who tested false-positive had a 15.5 percent chance of reprieve. In places that enjoy a low incidence of the disease, the same test would be even more likely to show true negatives as positive. In North America, for example, the incidence of AIDS is still well less than 1 percent, but I will take it as 1 percent for the sake of simplicity. Here again, the street math strategy of using simplifying assumptions makes the calculation much easier.

This time, in a hypothetical population of 1,000 randomly selected North Americans, 1 percent, or just 10, have AIDS. As I explained earlier, this level is still much higher than the actual incidence and I use it only as an illustration. Right away, if 10 people have AIDS, 990 do not have AIDS. If a testing center employs the same, 98 percent accurate test mentioned in Chapter 8, it will show 2 percent of the non-AIDS population (19.8 people of the 990) as positive, and it will also show 98 percent of the AIDS population (9.8 people of the 10) as positive. One thing that turns some people off mathematical language and mathematics generally is the use of fractional objects (not to mention people) in statistical reports of one kind or another. It's all right to use such figures in a calculation, but not in a final report, let alone a newspaper story. The calculator, after all, knows that "9.8 people" really stands for a proportion, "9.8 people of every 10" in this case.

Meanwhile, if a total of 19.8 + 9.8 = 29.6 people test positive, we may as well go on accepting fractional people and recognize that the proportion of cases that test positive for AIDS but are actually negative leaps to 19.8 out of 29.6, or almost 67 percent. The person who tests positive now has a much higher probability of reprieve, namely .662 as opposed to the .155

probability in the earlier AIDS-heavy population. Cheerfully enough, those who test negative have a much lower chance of being actually positive in the 1 percent population than in the 10 percent population, as readers are free to discover for themselves. Readers are also free to repeat these calculations for the actual AIDS test, which has the much higher accuracy of 99.997 percent.

In chancy situations where you have money on the line, there is a most useful general idea that we would all do well to carry in our heads. It is called *expected value* and, in most street examples, it follows a simple pattern: If there's a probability P and two amounts of money, one connected with P and the other with its opposite, $1 - P$, you must multiply each amount by its probability and add the numbers up. In the lottery example, I pointed out that the expected value of the player's winnings made the lottery a "sucker's bet."

The lottery player has a .000000013 probability of winning and a .999999987 probability of losing. Multiply these probabilities by the player's loss or gain in each case and add the resulting numbers together. But be sure to reflect loss with a minus sign and gain with a plus:

$$.000000013 \times 999{,}999 - .999999987 \times 1$$

In the first place, if the player wins, the return is the $1,000,000 prize minus the dollar that he or she paid for the ticket, a total of $999,999. In the second place, if the player loses, the return is a negative dollar, so to speak. When you multiply these amounts by their corresponding probabilities and add them, as above, you get –.987, an expected loss of 98.7 cents for every dollar you spend.

Expected value calculations generally make more sense to people when they see the results of a calculation come true over a long period of time. For example, in the case of the "flip-out" game mentioned in Chapter 4, you may calculate the expected value of the gambler's winnings in a single cycle of play. If the bet is $10 and the probability of winning is .5, then the gambler wins $10 or loses $10, depending on the outcome of the coin toss. Using exactly the same template:

$$\text{gain} \times \text{probability of gain} - \text{loss} \times \text{probability of loss},$$

you will get $10 \times .5 - 10 \times .5 = 0$ as the gambler's expected winnings. In the short run, he or she either loses $10 or wins $10. In the long run, the ratio of wins to total plays will tend to 1/2 and the gambler will lose $10 about as frequently as he or she wins that amount. He or she stands to make nothing in the long run. Strictly speaking, this formula only applies as long as the gambler has money to bet. If he or she runs out, he or she will be faced with gambler's ruin.

In the long run, the same gambler will also experience swings away from the zero-balance point, some containing runs of heads or tails. The runs can be arbitrarily long, and each length of run has a definite probability associated with it. As long as the gambler has a finite amount of capital, say $1000, he or she runs the risk of losing it all. If the bet is $10, for example, and he or she bets on heads and doubles the bet each time he or she loses, a run of seven tails will ruin him or her. By the multiplier rule, the probability of seven consecutive tails is $(.5)^7$ or .0078. Sooner or later it will happen.

With this understanding of expected values, you will appreciate that the husband and wife team who were offered stereo insurance for $20 (Chapter 6) did not consider it a good bet if 1/4 of the units were returned by customers. If they bought the insurance, their total, one-time cost would be $60 + $20, or $80. But if they did not buy the insurance, their expected cost would be

$$120 \times .25 + 60 \times .75$$

This calculation makes sense if you think about 100 couples, all deciding not to buy insurance. In 25 percent of the cases, the stereos will fail and those couples will be out $120, the price of two stereos. But in 75 percent of the cases, the stereos will work just fine and those couples will be out only $60, the price of one stereo. Since couples will be losing money in either case, the expected amounts are added together. The sum paid by the average couple would be $75, which is $5 less than the cost of the stereo with insurance. The couple in question used betting statistics to minimize their cost. Whenever professional gamblers find a situation like this, with a positive expected payoff, they *jump* at the chance to play.

Combinatorics Counts

Combinatorics is a wonderful branch of mathematics. Not only are its elements easily learned by nonmathematicians, but they are tremendously useful in calculating probabilities, distributions, and all kinds of useful and interesting facts. This is because, as I pointed out in the lottery example, combinatorics counts.

Combinatorics has a definite place in street math when it comes to figuring out how many times certain events can happen. Take the case of the mutual funds that had a 1/2 probability of outperforming the market in any one year. How many funds out of 1,000, say, would you expect to outperform the market eight years out of ten?

The answer depends on the number of ways that you can flip a coin ten times and come up with eight heads. You could start by enumerating the possibilities by writing down all the patterns of *H* (for heads) and *T* (for tails):

HTHHHTHHHH, THHHHHHTHH, ...

and so on. Inspecting the patterns, however, you would discover that it really comes down to the number of ways you can insert two Ts into a string of ten places. There are ten places you can put one T and, for each of these, nine places you can put the other. The total number of ways is therefore $10 \times 9 = 90$. As in the combinatorics of the 6-49 lottery, we have overcounted here, as well. Each pair of Ts gets counted twice by this formula, once when the "one" T comes before the "other" in the sequence and once when it comes after. The true answer is, therefore, $90/2 = 45$.

Since the total number of possibilities for the ten-coin toss is 2^{10}, or 1,024, the probability of getting exactly eight heads is simply:

$$\frac{45}{1024}$$

This probability equals .0439 or 4.39 percent. If all of the 277 funds mentioned in Chapter 2 were purely random in their tendencies to outperform the market, you would expect that 4.39 percent of them, about 12, would outperform the market eight years out of ten.

In this final chapter, I have not only shown how some of the answers used in earlier chapters were arrived at, but I have also illustrated two of the most powerful tools of streetwise numeracy: the back-of-the-envelope calculation and the deployment of general principles.

It remains only to remark, as far as general principles are concerned, that they occur on all levels. On the one hand, the principle of probability as a ratio of favorable outcomes to all possible outcomes may be applied anywhere that probability is relevant. On the other hand, the single most important principle of algebra, replacing an unknown number by *x*, applies even more broadly as several of the foregoing examples have shown.

I am encouraged to imagine that some readers will not only become more numerate with the help of this book but that they will discover, as I once did, the magic behind that one creative act: "Let the answer be *X*." How strange that a bold act of pretension (that you already know the answer as *X*) leads either sooner or later to a value for *X*, the answer itself!

Annotated Bibliography

The following books amount to a basic library for the numerate. Some analyze math abuse, some discuss the problems of innumeracy, and some illustrate the role that mathematics plays in the areas covered by *Two Hundred Percent of Nothing*. Other books involve general reading of proven worth and popularity. The books are arranged alphabetically by author, each accompanied by a minireview on which readers may, hopefully, base a buying decision.

Thomas A. Bass, *The Eudaemonic Pie* (Boston: Houghton Mifflin, 1985).

A West Coast commune of physicists decides to win big at roulette by building a computer that will predict which number the ball will land on. To avoid detection, the scientists fit the computer into the heel of a shoe. After much trial and error and abject failure at Las Vegas followed by new technical triumphs at the drawing board, the group is ready for the big kill. *The Eudaemonic Pie* (which refers to a hypothetical pie that the group will split) is filled with interesting observations about gambling and gamblers. What happens when mathematics meets the mafia?

Robin M. Dawes, *Rational Choice in an Uncertain World* (San Diego, CA: Harcourt Brace Jovanovich, 1988).

If thinking is like swimming, we can all learn to think better by bringing more control to our thought processes. Like the swimmer who learns to overmaster instincts by assuming a horizontal position in the water instead of a vertical one, a rational thinker learns to go against instinct from time to time. This book surveys dozens of common thinking patterns that are inappropriately deployed in circumstances that range from buying dresses to checking insurance. Backed by an amazing amount of research in the social sciences, Dawes shows us the cherished myths and bad cognitive habits we must give up, then explains how probability and statistics come to our rescue in an uncertain world.

Thomas Gilovich, *How We Know What Isn't So: The Fallibility of Human Reason in Everyday Life* (New York: The Free Press, 1991).

As Artemus Ward once said, "It ain't so much the things we don't know that get us into trouble. It's the things we know that just ain't so." Gilovich explains how people continue to believe things that "ain't so," even in the face of evidence to the contrary. More people believe in ESP than evolution, to cite just one example. Gilovich argues that just as our perceptual apparatus unavoidably makes us subject to various optical illusions, so our cognitive facilities ineluctably draw us into mistaken beliefs—especially when the world around us seems to reinforce the belief. The book contains illuminating discussions of how we see patterns in random data, how we are influenced by internal biases, and even how social consensus forces us to know things that ain't so.

Darrell Huff, *How to Lie with Statistics* (New York: W. W. Norton, 1982).

There is no better introduction to statistical innumeracy and abuse. Huff's humorous and straightforward discussion of such major topics as sample bias, abused averages, empty figures, missing numbers and gee-whiz graphs weave generic examples and mathematical rescue work together into a handy, pocket-sized book, just right for the street. Consistent cartoonage adds a light, graphic touch to illustrate every important concept. The perfect companion to the statistical side of this volume, *How to Lie with Statistics* pokes gentle but deadly serious fun at the mental manipulators that surround us.

A. J. Jaffe and Herbert F. Spirer. *Misused Statistics: Straight Talk for Twisted Numbers* (New York: Marcel Dekker, 1987).

Jaffe and Spirer trace the misuse of statistics from the quality of basic data that goes into a statistical study through methodology and on to interpretation of the results. They show how what *can* go wrong *does* go wrong in a number of fields, including some applications of paramount social importance: polls, affirmative action, and federal government reports. The examples are sometimes fictitious and sometimes real, but they always carry the points. The book begins with an example of an ideal statistical study in the service of humanity, namely Florence Nightingale's careful analysis of wound recovery that finally convinced the British Army to introduce sterile methods into field hospitals. The book ends with an amusing discussion of *ectoplastistics*, a mouthful meant to convey a number that has no real meaning and that could only have come from a séance.

Jeffrey Katzer, Kenneth H. Cook, and Wayne W. Crouch. *Evaluating Information: A Guide for Users of Social Science Research,* 2nd ed. (New York: Random House, 1982).

This book not only should be read by members of the media, it *could* be read by them! From the introduction: "Much information reported by scientists, published in reputable journals, and used by students, practicing professionals, and the general public is misleading. Some of it is just plain wrong. The purpose of this book is to help you detect such misinformation." Intended for readers of research, *Evaluating Information* helps them detect many of the common errors in published research. It covers the difficulties that even scientists have in observing their results, the errors that arise in communicating their research, the errors that readers commonly make in understanding it, and the errors that arise from misunderstanding methodology and its limitations. Written in a common-sense style and accessible to lay readers, this book should be on the shelf of everyone with a stake in understanding scientific research and the way it is reported.

Burton G. Malkiel, *A Random Walk Down Wall Street,* 4th ed. (New York: W. W. Norton, 1985).

Malkiel, dean of the Yale School of Organization and Management, takes a clear-eyed, humorous but basically unsympathetic look at the illusions and delusions of the investment industry. He builds a convincing case that technical analysis has about the same chance of predicting stock prices as reading chicken entrails would. He even attacks fundamental analysis as without hope in a world where the madness of greed and fear drive prices of poor stocks through the roof and prices of good stocks into the ground. This engaging read ends with hard-nosed advice for individual investors.

David W. Moore, *The Super Pollsters* (New York: Four Walls Eight Windows, 1992).

Setting the stage by detailing Shere Hite's fall from statistical grace (following publication of her books that claimed most women get little or nothing out of sex), Moore traces the history of public polling, from George Gallup in the 1930s to Mervin Field and his California Poll in the 1940s. Both polls continue to this day, having survived the vicissitudes of new techniques and old bugbears: sample bias, sampling techniques, the effect of questions, interpretive problems, and so on. The subtleties of polling, the interplay of its political and mathematical dimensions, comprise the major threads of the book's fabric. Holding it all together are the fascinating personalities of the pollsters themselves, the politicians and public figures they dealt with, and the many behind-the-scenes stories that give new insights into the measurement and manipulation of public opinion in the United States.

James R. Newman, *The World of Mathematics* (New York: Microsoft Press, 1988).

Subtitled "A small library of the literature of mathematics from A'h-mose the scribe to Albert Einstein," this four-volume set truly is a library. Never have so many good writers and experts congregated in a single place to provide a myriad of views of mathematics, both ancient and modern. The volumes march grandly through historical and biographical material, numbers and the art of counting, space and motion, mathematics and the physical world, mathematics and social science, the laws of chance, statistics, and the logic of mathematics. We find mathematics applied in warfare, art, music, and even ethics. The fourth volume rounds off with a host of amusements, puzzles, and fancies. Were it not hyperbole, one could say that countless readers have found themselves drawn back countless times to the timeless charm, incisive thought, and eminent readability of this great classic.

John Allen Paulos, *Innumeracy: Mathematical Illiteracy and Its Consequences.* (New York: Vintage Books, 1990).

Innumeracy, a book that raised the mathematical consciousness of hundreds of thousands of people, introduces a wide variety of mathematical ideas written in an imaginative, engaging, and sometimes peculiar style: "If walls were built about it [Central Park], all the blood in the world would cover the park to a depth of something under 20 feet." Paulos' examples, meant to illuminate mathematical ideas, rarely smell of the street in spite of their deliberately familiar settings. They are manufactured from logical cloth or adapted from the "folklore" of mathematics itself. For people who enjoy *The Skeptical Inquirer* and want to see some junk-science bashing, from Freud to the horoscope, Paulos obliges with a well-explained tour of science abuse.

(Senator) William Proxmire, *The Fleecing of America* (Boston: Houghton Mifflin, 1980).

Senator Proxmire established the Golden Fleece Award in 1975 to highlight wasteful spending by the U.S. government. Proxmire documents misspending by the government in grants to universities (to investigate Peruvian brothels and to fund an artist to drop burning paper from the sky, among other things), military misspending (the navy sends 64 aircraft and 1,334 reserve officers to a social function in Las Vegas), Washington waste (ferrying low-level functionaries around in hundreds of limousines), public works rip-offs (fourfold cost overruns on government luxury buildings), and aid to the poor (which ends up in the pockets of the rich), to name a few. Proxmire looks into the reasons that governments overspend. He analyzes the power of lobby groups, from bankers to builders. He reports on two success stories in controlling spending and ends with suggestions for controlling the annual budget deficit.

John Scarne, *Scarne's Guide to Casino Gambling* (New York: Simon and Schuster, 1978).

This is a compendium of casino games, each carefully described and illustrated by colorful prose and tales of the high rollers, from Las Vegas to London. Some monstrous wins are balanced by many tales of woe and a sober look at house percentages for blackjack, bank craps, roulette, slot machines, baccarat, the money wheel, and others. Scarne loses some grace by overplaying his own importance in the history of gambling and loses a bit more when, after a useful and realistic discussion of probability in gambling, he falls for the hot-streak theory.

Richard H. Schwartz, *Mathematics and Global Survival*, 2nd ed. (Needham Heights, MA: Ginn Press, 1991).

Designed for the classroom, this book motivates students through examples that highlight the dangers to the environment, from warfare to human pollution. Introducing concepts and operations in easy stepwise fashion, the book never loses its focus on one aspect or another of the environment. Here are just some of the topics: growth of populations, survival of species, pollution versus resource usage, trees and newspapers, worldwide energy use. It would be hard to find a relevant calculation that doesn't appear in Schwartz's book. Great for the general public, too. Look up your favorite environmental issue. If you don't find it, you will at least find techniques, from graphical methods to statistics, for describing and analyzing it.

Lynn Arthur Steen (Ed.), *On the Shoulders of Giants: New Approaches to Numeracy* (Washington, D.C.: National Academy Press, 1990).

Here is a book rich with inspiration for those who want to find out what mathematics and mathematical thinking is all about. Steen gathers essays from expert writers on the subjects of dimension (Thomas F. Banchoff), quantity (James T. Fey), uncertainty (David S. Moore), shape (Marjorie Senechal), and change (Ian Stewart). Each essay shows mathematics in action in the world around us. Each essay also sketches the mental skills that we need to hone in our students. Mathematics has changed as the world has changed. New subjects such as dynamics and chaos and new technology such as computers challenge us to frame a curriculum that will be relevant in the twenty-first century. Here are not only the key ideas but indications of the key skills we must develop in ourselves and our students.

Lester Thurow, *Head to Head: The Coming Economic Battle Among Japan, Europe and America* (New York: William Morrow, 1992).

Who will win the competition for the production of new goods and services in the 1990s and beyond? Europe, Japan, or America? Thurow places America third in a race it cannot afford to lose. Europe and, especially, Japan seem willing to forego short-term profits for reinvestment in plant, development, and research. The longest term investment of all, education, will be the major, indispensable foundation of success. Even as Europe and Japan improve their educational systems, the American system is unraveling thanks to absent educational standards and an ingrained belief that nothing is wrong.

Sheila Tobias, *Overcoming Math Anxiety* (Boston: Houghton Mifflin, 1978).

Nobody has written a better analysis of the reasons and remedies for math anxiety than Sheila Tobias. Math anxiety begins for many people with a sense of failure, a sense of being shut out from special insights that seem to require a special kind of mind. But Tobias shows how seemingly incurable people may be brought to the "aha!" insight that marks the beginning of real interest in mathematics and a sudden feeling that it isn't so difficult after all. She cuts through the cultural and sexual stereotypes that have kept multitudes of people away from the feast. Throughout the book, examples not only illustrate the difficulties people have in solving mathematical problems, even of the simplest type, but they provide a kind of surreptitious education in the subject. There is still no better medicine for the math-anxious.

Edward R. Tufte, *The Visual Display of Quantitative Information* (Cheshire, CT: Graphic Press, 1983).

This large-format book, widely considered a classic, considers both good and bad examples of how numerical or quantitative information is best presented to readers eager to capture the gist of numbers. Fascinating maps and charts in the section on excellence show the prevalence of stomach cancer in the United States, ocean currents, trade routes, and galactic images. It amazes one how much meaningful information can be packed into a well-designed graphic display. But "data ink" and "chart junk" make their appearance in a second section. Here are weird and wonderful examples of mathematical and graphic imaginations gone wild, with subsequent confusion for innocent eyes and great amusement for abuse detectives. Tufte's book is almost a course in visual mathematics and a "must" read for anyone involved in the graphics arts, technical writing, or publishing business.

Michael Wheeler, *Lies, Damn Lies, and Statistics. The Manipulation of Public Opinion in America* (New York: Dell, 1976).

President Nixon, who tried to manipulate pollster's results, has departed from the political scene, but the pollsters are still with us, argues Wheeler. The consistently low popularity that the Harris poll attributed to Nixon may have had its origin in the kind of question that the Harris pollsters asked respondents. Equipped with reams of background information on some of the most important developments in U.S. politics, Wheeler walks the reader through the history of polling from the Landon-Dewey struggle right up to the post-Nixon years of the late 1970s. The book includes key chapters on the effect of polls on the Vietnam struggle as well as an absorbing account of the Nielsen rating system, how it works, and the distorting effect it has on television content.

Mathematical Sciences Education Board, National Academy of Sciences, *Everybody Counts: A Report to the Nation on the Future of Mathematics Education* (Washington, D.C.: National Academy Press, 1989).

The Mathematical Sciences Education Board, comprised of 70 leading teachers, professors, researchers, engineers, and administrators across the United States, has assembled a preliminary report, the first in an ongoing series that will keep the U.S. public abreast of the latest plans and developments in mathematics education. The report examines all grades, from kindergarten through to university graduate school, and all aspects of the educational process, from curriculum to human resources. The report documents changing trends in enrollment demographics, outlining new challenges that face today's and tomorrow's teachers. It outlines the increasing importance of mathematics in today's society, the philosophy and standards that must stand behind revised curricula, new approaches to teaching such as learning by involvement, and general goals to be met by the twenty-first century educational system. The report takes a generic, apple-pie view of the challenges that face mathematics education and makes very few specific recommendations.

Acknowledgments: The Abuse Detectives

This book would not have been possible without the examples of math abuse sent in by hundreds of people across North America and, for that matter, around the world. I owe special thanks to the following individuals whose examples were used in the foregoing pages:

Drew Auth, Tom Bach, William D. Barclay, George Barry, Raymond J. Bayerl, Carter Bays, Russell Bell, Russell R. Bergquist, Aaron C. Brown, Homer E. Brown, Joseph Childers, L. E. Cannon, Paul Ciernia, Fred D. Clark, David E. Cochran, Andy Condon, Harry C. Crim, Jr., Ross Daily, Gehl Davis, Cecil Deisch, Alexander Denton, H. Harvey Dewing, Tom Dolan, Donald DuBois, Paul C. Eklof, Joseph Erker, Bertrand Fry, Ray Galt, Peter Gordon, George W. Greenwood, Dale E. Hammerschmidt, Peter Hardie, John Hart, Richard E. Haswell, Hans Haverman, Richard L. Henschel, Edward F. Hogan, C. E. Holvenstot, Shafi U. Hossain, J. E. Householder, Doug Jackson, Roger H. James, Karl Johanson, Frithiof V. Johnson, Amy L. Justus, Myron B. Katz, David M. Koppelman, Paul Ladanyi, Jane I. Lataille, Sanford H. Lefkowitz, Alan L. Lehman, J. Samuel Listiak, Jan M. Ludwinski, Robert M. Martin, Gus Mancuso, Howard Mark, Alexander Martschenko, Jacob E. Mendelssohn, V. Mezl, Ruben Misrahi, Doug Mitchell, Robert K. Multer, Tom Napier, William

Nighman, Mark H. North, Frank Palmer, Keith R. Park, Paul Printz, Donald I. Promish, George E. Reeves, Anthony Riddle, Carl H. Savit, R. C. H. Schmidt, Pedro Seidemann, Raymond C. Shreckengost, Daniel Snowden-Ifft, David R. Swenson, Arthur Summers, David A. West, Michael Wheeler, L. W. Whitlow.

Be a Math Abuse Detective

As long as innumeracy continues at its current high levels, math abuse will be with us. This will make a future update necessary. Readers who know of specific abuses are urged to send them to the author in care of: "200 Percent," John Wiley & Sons, Inc., 605 Third Avenue, New York, NY 10158.

Further Acknowledgments

I owe further thanks to my agent, Linda McKnight, to Wiley editors Steve Ross (for getting the book started) and Emily Loose (for keeping the prose on target). Thanks also to Kathleen McGlaughlin for research and to my colleague Professor Seymour Ditor for critical readings.